FUN With CALCULUS

Easy to use, Easy to Learn

500 Full Solutions and Examples

Marcel Sincraian, Ph.D.

ISBN: 978-1-7775022-1-8 Electronic Book

ISBN: 978-1-7775022-0-1 Printed Book

Marcel Sincraian Email: msincraian@yahoo.ca

To:

My mom and dad

To:

My loving wife and precious daughters

To:

My sister

To:

All my hard-working students that kept me on my toes:

To:

Melissa, Luke, Emma C, Adrian, Ben

Content

Chapter 1 Functions 1

A. Parent functions (Review Pre-Calculus 11 and 12) 3
- a. Determine: if a relation is a function, the values of a function, the range 3
- b. Linear and quadratic functions and their graphs 5
- c. Inverse functions and their graphs 7

B. Piecewise functions 9
C. Trigonometric functions 11
D. Graphs of trigonometric functions 14
- a. Graphing sine and cosine functions 14
- b. Graphing tangent and cotangent functions 16

E. Inverse trigonometric functions 19
F. Graphs of inverse trigonometric functions 21

Quick Answers 24

Full Solutions 25

Chapter 2 Limits 39

A. From table of values, graphically, and algebraically 43
- a. Table of values, and graphically 43
- b. Algebraically 44

B. One side versus two sided 47
C. End behavior 49
D. Intermediate value theorem 51
E. Left and right limits 53
F. Limits to infinity 55
G. Continuity 58

Quick Answers 61

Full Solutions 62

Chapter 3 Differentiation 75

A. History 77
B. Definition of derivatives 78
C. Notation 80
D. Rate of Change 81
- a. Average versus Instantaneous 81
- b. Slope of secant and tangent lines 83

E. Transcendental functions 85
- a. Logarithmic, Exponential, Trigonometric 85

F. Differentiation rules 87
a. Power 87
b. Product 89
c. Quotient 91
d. Chain 93
G. Higher order differentiation 95
H. Implicit differentiation 97
I. Applications 99
a. Relating graph of f(x) to f '(x) and f ''(x) 99
b. Differentiability, mean value theorem 101
c. Newton's method 103
d. Problems in contextual situations, including related rates and optimization problems 106
Quick Answers 108
Full Solutions 109

Chapter 4 Integration 149

A. Definition of an integral and notation 151
B. Definite and indefinite integrals 153
C. Approximations 155
a. Riemann sum, rectangle approximation method, trapezoidal method 155
D. Fundamental Theorem of Calculus 158
E. Methods of integrations 161
a. Antiderivatives of functions 161
b. Substitution 163
c. By parts 165
F. Applications 167
a. Aria under a curve, volume of solids, average value of functions 167
b. Differential equations Initial value problems and slope fields 172
Quick Answers 174
Full Solutions 175
Pictures used 200
Formulas 202
Appendix 204
Bibliography 206

Introduction

It is generally known that for some students, Calculus might be a hard subject. This book provides students with a tool to improve their knowledge in Calculus; this is done in a light hearted manner in order to help students having fun while practicing Calculus. The questions in this book come from general knowledge regarding different fields, such as: the Roman Empire, cars, animals, mountains, Apollo missions to the Moon, and architecture. This, in turn, will refresh some of the knowledge the students acquired in Science and Social Studies, and hopefully make studying easier and fun.

This book is intended to be a quick review of the Calculus course. It includes the main chapters within the Calculus course. The book follows the Calculus I curriculum for high schools. Chapter 1 is a review of Functions; Chapter 2 is about Limits; Chapter 3 deals with Differentiation, and Chapter 4 is all about Integrals.

How to use this book?

Before each set of practice problems, a short review of the main theoretical concepts will be presented. There are examples given to help better understand and review the concepts.

The practice problems page begins by presenting a question to be answered. Students should read the question, then try all the problems. The students should follow the request of the question.

On the left side of the question, there will always be a picture that symbolizes the subject area the question comes from like Roman Empire, architecture, etc.

Each practice problem has a number from 1 to 10 and it is assigned a certain letter.

There will be some practice problems that are not the correct letter. The students should solve all the problems in order to find the correct result.

The letters of the problems that are correct, are to be crossed out in the table provided at the bottom of the page. The letters that are not crossed out will represent the word or abbreviation of a word that is the answer to the question at the top of the page.

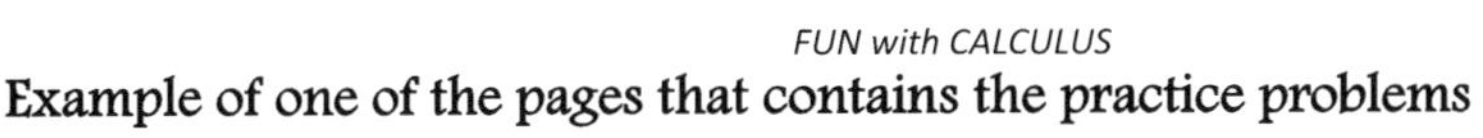

Example of one of the pages that contains the practice problems

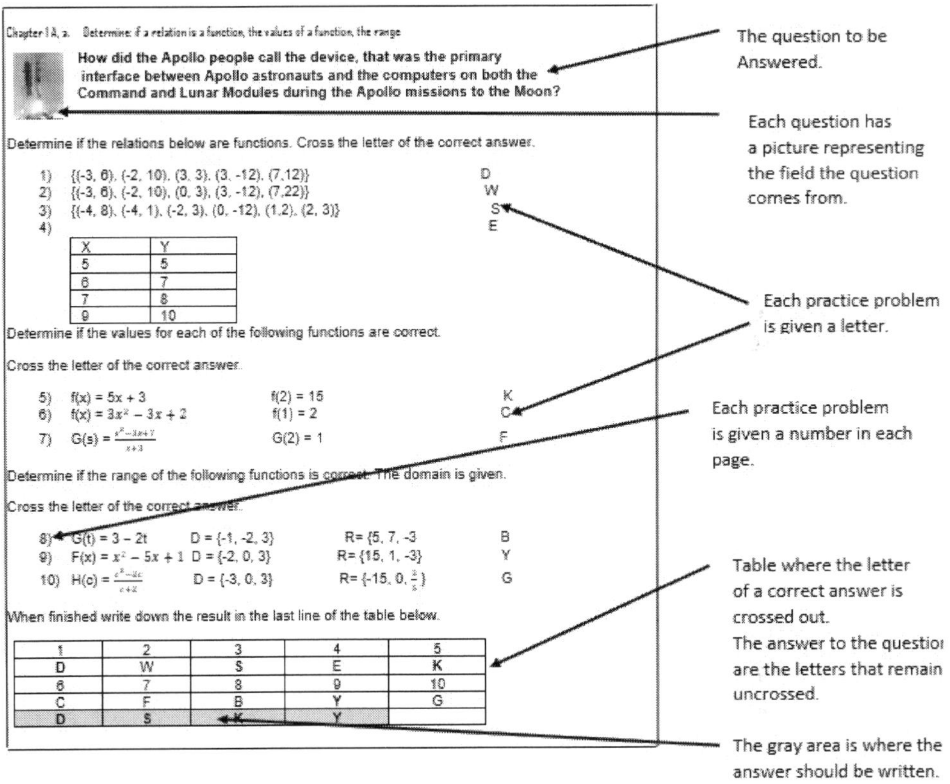

Chapter 1 A, a. Determine: if a relation is a function, the values of a function, the range

How did the Apollo people call the device, that was the primary interface between Apollo astronauts and the computers on both the Command and Lunar Modules during the Apollo missions to the Moon?

Determine if the relations below are functions. Cross the letter of the correct answer.

1) {(-3, 6), (-2, 10), (3, 3), (3, -12), (7,12)} D
2) {(-3, 6), (-2, 10), (0, 3), (3, -12), (7,22)} W
3) {(-4, 8), (-4, 1), (-2, 3), (0, -12), (1,2), (2, 3)} S
4) E

X	Y
5	5
6	7
7	8
9	10

Determine if the values for each of the following functions are correct.

Cross the letter of the correct answer.

5) $f(x) = 5x + 3$ $f(2) = 15$ K
6) $f(x) = 3x^2 - 3x + 2$ $f(1) = 2$ C
7) $G(s) = \frac{s^2-3s+7}{s+3}$ $G(2) = 1$ F

Determine if the range of the following functions is correct. The domain is given.

Cross the letter of the correct answer.

8) $G(t) = 3 - 2t$ D = {-1, -2, 3} R= {5, 7, -3 B
9) $F(x) = x^2 - 5x + 1$ D = {-2, 0, 3} R= {15, 1, -3} Y
10) $H(c) = \frac{c^2-4c}{c+4}$ D = {-3, 0, 3} R= {-15, 0, [illegible]} G

When finished write down the result in the last line of the table below.

1	2	3	4	5
D	W	S	E	K
6	7	8	9	10
C	F	B	Y	G
D	S	K	Y	

The question to be Answered.

Each question has a picture representing the field the question comes from.

Each practice problem is given a letter.

Each practice problem is given a number in each page.

Table where the letter of a correct answer is crossed out.
The answer to the questio[n] are the letters that remain uncrossed.

The gray area is where the answer should be written.

Each chapter begins by listing all the questions to be answered.

Solutions

At the end of each chapter, there are quick answers with the words that students should obtain from solving the questions. These words represent the correct result of the page.

A Full Solution it is provided at the end of each chapter, that explains all the problems of the book step by step.

Notations

$f(x)$ - function of x

$f^{-1}(x)$ - inverse function

$\lim_{x \to a} f(x) = L$ - the limit L of a function f(x) for a value of x approaching a value "a"

$\frac{\Delta y}{\Delta x}$ - rate of change, or slope

$f'(x)$ - first derivative of function f(x)

y' - first derivative,

$y'' = f''(x)$ – second derivative

$\int f(x)dx$ – indefinite integral of the function f(x)

$\int_a^b f(x)dx$ - definite integral of the function f(x) for x between value a and value b

F(x) is an antiderivative of f(x) if $F'(x) = f(x)$

$\sum_{i=1}^{n} x_i$ – the sum of all values x_i for i being a natural number between 1 and n

X^2 - x^2 the sign ^ represents the fact that the number after it is an exponent

X^n - x^n

System of axes: Horizontal is x axis; Vertical is y axis.

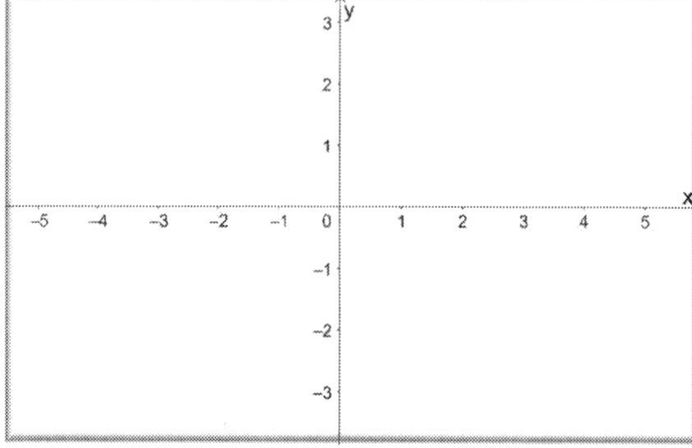

CHAPTER 1

Functions

"Give me but a firm spot on which to stand
and I shall move the earth."

Archimedes (c. 287 – c. 212 BC)

(Greatest mathematician of Antiquity)

Questions to be answered:

- What did the people working on the Apollo missions call the device that was the primary interface between Apollo astronauts and the computers on both the Command and Lunar Modules during the Apollo missions to the Moon?
- The first triumvirate created by Gaius Julius Caesar, Marcus Licinius Crassus, and Gnaeus Pompeius Magnus had the purpose of making sure they ruled Rome. Gnaeus Pompeius Magnus collaborated with Ceasar first, then became his greatest foe. What was his nickname?
- In 1913, what company made the first mass-produced automobile?
- In what country was the route that Edmund Hillary and Tenzing Norgay took in May 1953 on the first ever ascent of Mount Everest (8,848m)?
- How long are the legs of a foal compared with a mature horse? They are approximately.......
- Who was one of the most persistent enemies of Gaius Julius Caesar whom, after a life time of opposing Caesar in the Roman senate as member of "optimates" and always Caesar's equal in persistence, and force of character, died in Africa at Utica (Tunisia) in 46 B.C. two years before Caesar was assassinated?
- What was the rocket that lifted the Apollo capsule called?
- What is the name of the mountain that is the tallest in the Canadian Rockies, is situated near Jasper and has a height of 3,954 m?
- What country in the world has the most cars?

1.A. Parent functions (Review Pre-Calculus 11 and 12)

a. Determine: if a relation is a function, the values of a function, the range

Theory and Examples

A relation is a function when for each value x that belongs to the domain, there is only one value y that belongs to the range.

EXAMPLE

Going through the function f in the picture below, there is only one value in the Range that corresponds to any value in the Domain. For value 2 in the Domain, it corresponds with only one value in the Range (4). For value 4 in the Domain, it corresponds with only one value in the Range (4). For value 3 in the Domain, it corresponds with only one value in the Range (9).

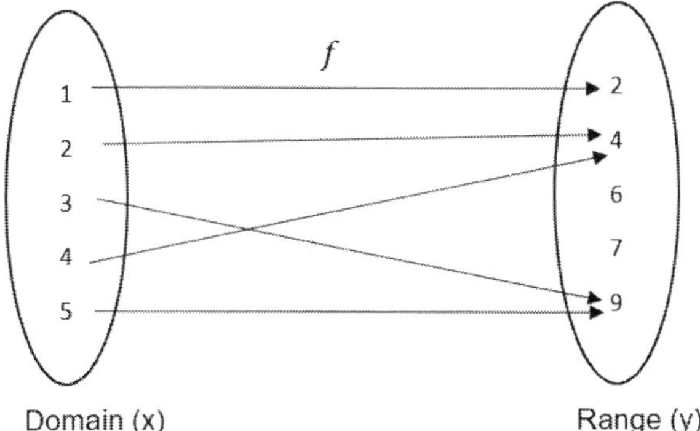

The value of a function is the value in the Range that is the result of a value in the Domain going through the function f.

EXAMPLE

In the figure above one value of the function f is 9. This number is the result of the domain value of 3 going through f as well as 5 going through f.
The Range is the set of elements (letters or numbers) that are the result of the elements (letters or numbers) in the Domain going through function f.

EXAMPLE

The set of numbers {2,4,6,7,9} in the figure above is the Range of f.
The Domain is the set of numbers {1,2,3,4,5}.

Chapter 1. A. a. Determine: if a relation is a function, the values of a function, the range

What did the people working on the Apollo missions call the device that was the primary interface between Apollo astronauts and the computers on both the Command and Lunar Modules during the Apollo missions to the Moon?

Determine if the relations below are functions. Cross the letter off for the correct answers. The uncrossed letters are the ones that make the answer to the question above.

1) {(-3, 6), (-2, 10), (3, 3), (3, -12), (7,12)}

2) {(-3, 6), (-2, 10), (0, 3), (3, -12), (7,22)}

3) {(-4, 8), (-4, 1), (-2, 3), (0, -12), (1,2), (2, 3)}

4)

X	Y
5	5
6	7
7	8
9	10

Determine if the values for each of the following functions are correct. Cross the letter of the correct answer.

5) f(x) = 5x + 3 f(2) =15

6) f(x) = $3x^2 - 3x + 2$ f(1) = 2

7) G(s) = $\frac{s^2-3s+7}{s+3}$ G(2) = 1

Determine if the range of the following functions is correct. The domain is given. Cross the letter of the correct answer.

8) G(t) = 3 – 2t D = {-1, -2, 3} R= {5, 7, -3}

9) F(x) = $x^2 - 5x + 1$ D = {-2, 0, 3} R= {15, 1, -3}

10) H(c) = $\frac{c^2-2c}{c+2}$ D = {-3, 0, 3} R= {-15, 0, $\frac{3}{5}$}

1	2	3	4	5
D	W	S	E	K
6	7	8	9	10
C	F	B	Y	G

Pronounced Diskey

1.A. Parent functions (Review Pre-Calculus 11 and 12)

b. Linear and quadratic functions and their graphs

Theory and Examples

A function is linear when the difference between consecutive values in the Domain is always the same. In the same time, the difference between consecutive values in the Range is always the same, not necessarily the same with the difference between consecutive Domain values.

EXAMPLE

Domain(X)	Range(Y)	Point
1	5	A
2	7	B
3	9	C
4	11	D

Here the difference between consecutive values in the Domain is 1. $x_B - x_A = 2 - 1 = 1$, or $x_D - x_C = 4 - 3 = 1$.
In the same time the difference between consecutive values in the Range is 2. $y_B - y_A = 7 - 5 = 2$, or $y_D - y_C = 11 - 9 = 2$.
When we graph a linear function, the graph is a straight line.

The pairs (x,y) represent points in the Cartesian system of axes.

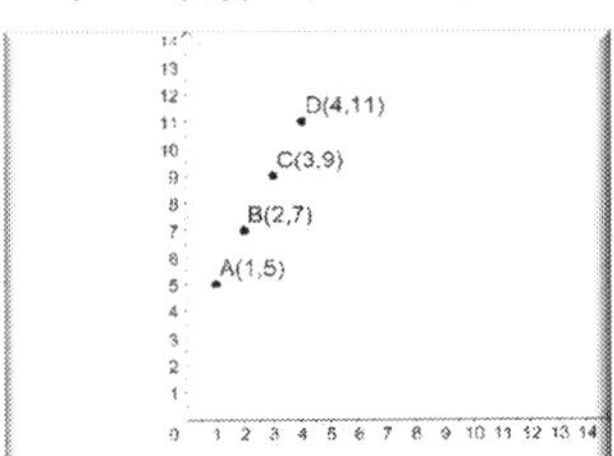

We can see here that if we connect these points it will create a straight line.
The equation that represents this graph is $y = 2x + 3$ and it is called a linear equation. (the grade of the equation is either 0 or 1).

A function is *quadratic* when the relation between x and y is a polynomial of second degree.
$y = x^2 - 5x + 4$ or $xy + 2x^2 - 3y = 5$

EXAMPLE

$y = x^2 - 5x + 4$

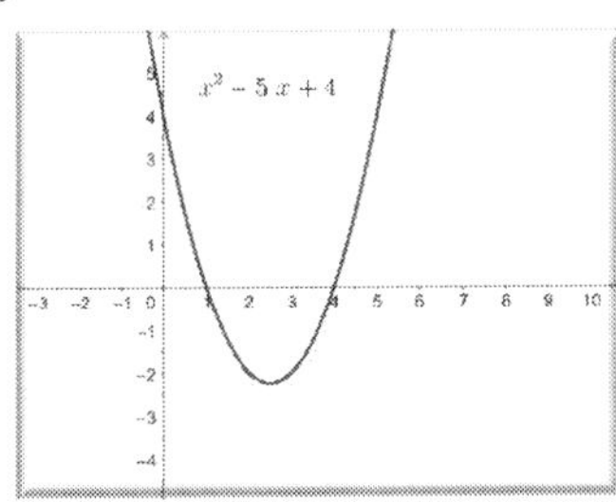

Here the quadratic function $y = x^2 - 5x + 4$ is represented.

VERTICAL LINE TEST

If we draw a vertical line through a graph and this line **intersects the graph** at **only one point**, then the graph represents a **function**.
If this line **intersects the graph** at **more than one point**, then the graph DOES NOT represent a Function.

Chapter 1. A. b. Linear and quadratic functions and their graphs

The first triumvirate created by Gaius Julius Caesar, Marcus Licinius Crassus, and Gnaeus Pompeius Magnus had the purpose of making sure they ruled Rome. Gnaeus Pompeius Magnus collaborated with Ceasar first, then became his greatest foe. What was his nickname?

Determine if the following tables of pairs of x and y and the expressions represent a linear function. In the table at the bottom of the page cross all the letters of the correct answer off.

1)

X	Y
5	5
6	7
6	8
9	10

2)

X	Y
1	3
2	5
3	7
4	9

3)

X	Y
2	0
2	5
3	10
4	15

4) $f(x) = 2x + 3$

5) $f(x) = x^2 - 3x + 4$

6) $f(x) = 34x - 15$

Determine if the following graphs represent a linear function. Cross all the letters for the correct answer off in the table at the bottom of the page.

7)

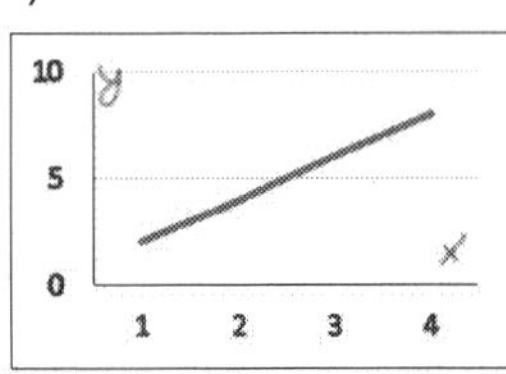

8)

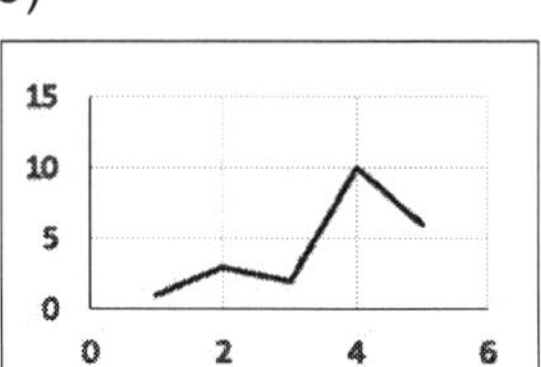

9)

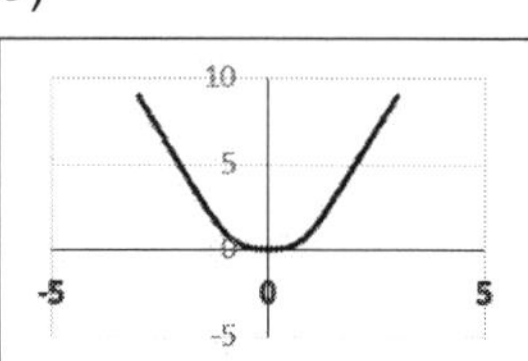

Determine if the following expressions represent a quadratic function. Cross all the letters for the correct answer off in the table at the bottom of the page.

10) $f(x) = 2x^2 - 3x + 4$ 11) $f(x) = x^2 - 3x$ 12) $f(x) = x^4 - 3x + 4$

1	2	3	4	5	6
P	R	O	E	M	A
7	8	9	10	11	12
Q	P	E	T	U	Y

1.A. Parent functions (Review Pre-Calculus 11 and 12)

c. Inverse functions and their graphs

Theory and Examples

An inverse function represented by $f^{-1}(x)$ is the function that has the Domain equal to the Range of the original function $f(x)$. The Range of the inverse function equals the Domain of the original function.

EXAMPLE

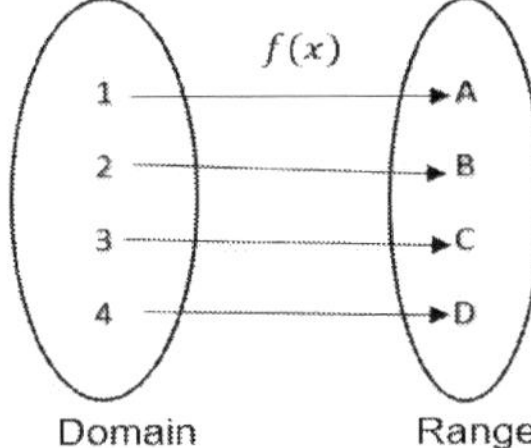

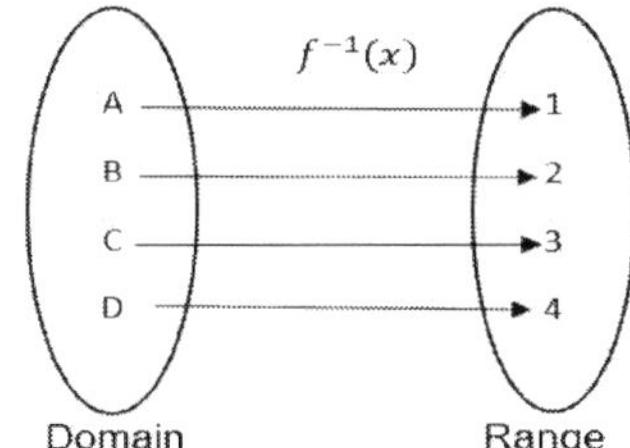

EXAMPLE

We obtain the equation of the inverse function in few steps:

1. Write the equation of the original function.

The original function is $f(x) = y = 4x + 7$

2. Switch the variables x and y in the original formula.

$x = 4y + 7$

3. Solve for y

$x - 7 = 4y$ so, $y = f^{-1}(x) = \frac{x-7}{4}$

The graph of an inverse function is always a reflection of the graph of the original function by y=x.

EXAMPLE

$f(x) = 4x + 1 \; and \; f^{-1}(x) = \frac{x-1}{4}$

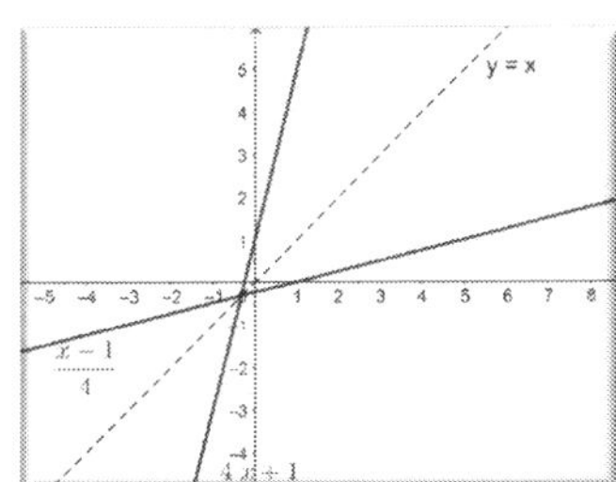

PROPERTY OF INVERSE FUNCTIONS

$$f\left(f^{-1}(x)\right) = f^{-1}\left(f(x)\right) = x$$

The function of the inverse function for a value x, equals the inverse function of the same function for value x, and equals value x.

Chapter 1. A. c. Inverse functions and their graphs

In 1913, what company produced the first mass-produced automobile?

Determine if the following expressions represent an inverse function. In the table at the bottom of the page cross all the letters of the correct answer off.

1) If $f(x) = y = 2x + 3$ the inverse function is: $f^{-1}(x) = \frac{x-3}{2}$

2) If $f(x) = y = 3x - 5$ the inverse function is: $f^{-1}(x) = \frac{x-5}{2}$

3) If $f(x) = y = x^2 - 1$ the inverse function is: $f^{-1}(x) = y = \frac{x-5}{2}$

4) If $f(x) = y = \frac{3}{2x+4}$ the inverse function is: $f^{-1}(x) = y = \frac{3-4x}{2x}$

5) If $f(x) = y = \frac{5}{2x^2+4}$ the inverse function is: $f^{-1}(x) = y = \frac{3-5x^2}{2x}$

6) If $f(x) = y = \frac{\sqrt{x-1}}{3}$ the inverse function is: $f^{-1}(x) = y = \frac{9-x}{2x^2}$

7) If $f(x) = y = \frac{3x}{x+3}$ the inverse function is: $f^{-1}(x) = \frac{3x}{3-x}$

8) If $f(x) = y = \frac{1}{5x+3}$ the inverse function is: $f^{-1}(x) = \frac{1-3x}{5x}$

Graph the following functions and their inverse functions on the same graph.

9) $f(x) = y = \frac{2}{3x+2}$ the inverse is: $f^{-1}(x) = \frac{2-2x}{3x}$

10) $f(x) = y = \frac{2x-1}{x+2}$ the inverse is: $f^{-1}(x) = \frac{-2x-1}{x-2}$

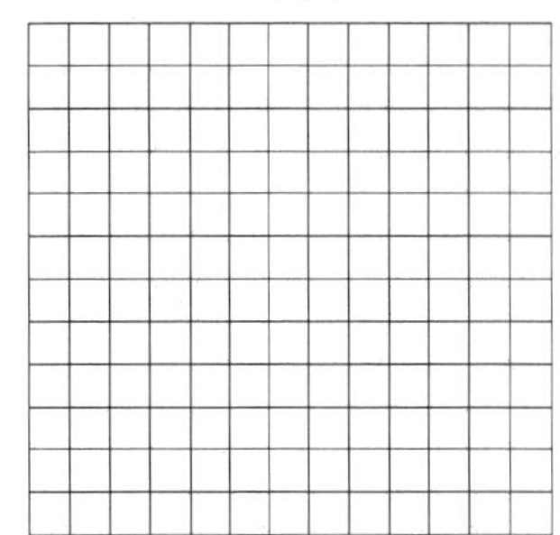

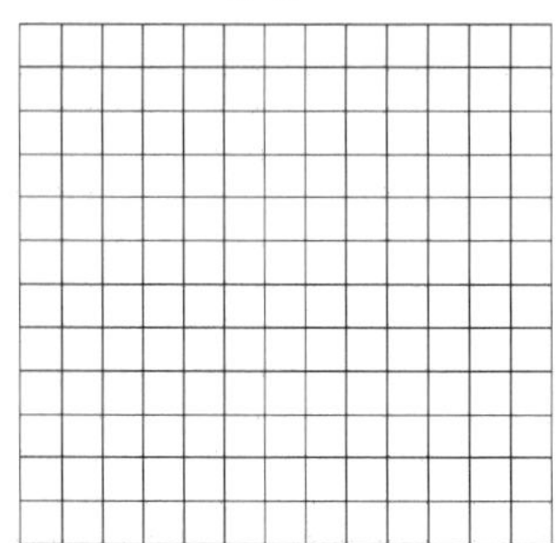

1	2	3	4	5	6	7	8
T	F	O	S	R	D	A	B

1.B. Piecewise functions

Theory and Examples

A *piecewise function* behaves differently on certain intervals of the Domain. It can be split into subsections.

EXAMPLE

$$f(x) = \begin{cases} x \; for \; x \leq -2 \\ 1 \; for -2 < x < 3 \\ 2x - 2 \; for \; x \geq 3 \end{cases}$$

This function has three subsections.

As we can see for all the values,

$x \leq -2 \; f(x) = x.$

For $-2 < x < 3$ the function is a horizontal line that crosses y axis at y=1.

For all the values of x greater or equal to 3, the function is a straight line with slope equal with 2 and y-intercept equal with -2.

EXAMPLE

$$f(x) = \begin{cases} 2 - x \; for \; x < 1 \\ 3 \; for \; x \geq 1 \end{cases}$$

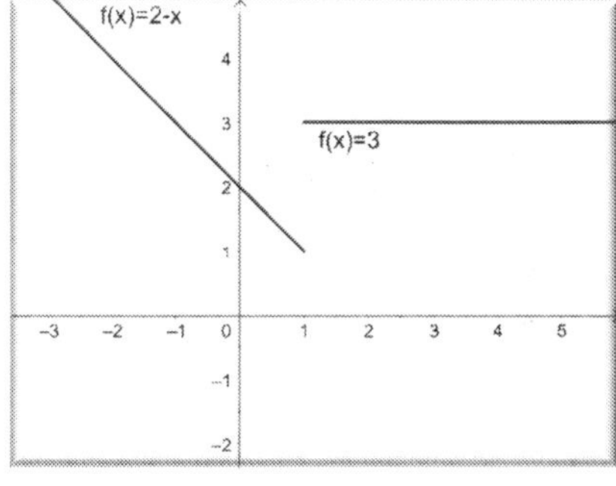

This function has two subsections.

As we can see for all the values,

$x \leq 1 \; f(x) = 2 - x.$

The straight line with slope of -1 crosses the y axis at y=2.

For all the values of x greater or equal to 1, the function is a straight line with slope equal to 2.

Chapter 1. B. Piecewise functions

In what country was the route that Edmund Hillary and Tensing Norgay took in May 1953 on the first ever ascent of Mount Everest (8,848m)?

Determine if the expressions below represent piecewise functions. In the table at the bottom of the page cross all the letters that are the correct answers off.

1) $f(x) = 2x^2 + 5x - 3\,, x \in R$

2) $f(x) = \begin{cases} 2x \; for \; x < 0 \\ 3 \; for \; x \geq 0 \end{cases}$

3) $f(x) = 4x + 3, x \in R$

4) $f(x) = 7x - 3, x \in R$

5) $f(x) = \begin{cases} 0.75x^4 \; for \; x < 2 \\ 3x + 6 \; for \; x \geq 2 \end{cases}$

6) $f(x) = 3\sin(x - 2) + 3, x \in R$

7) $f(x) = \begin{cases} x^2 \; for \; x \leq -2 \\ 4 \; for - 2 < x < 2 \\ 2x \; for \; x \geq 2 \end{cases}$

8) $f(x) = \sqrt{x - 1}$

Graph function 5 and 7.

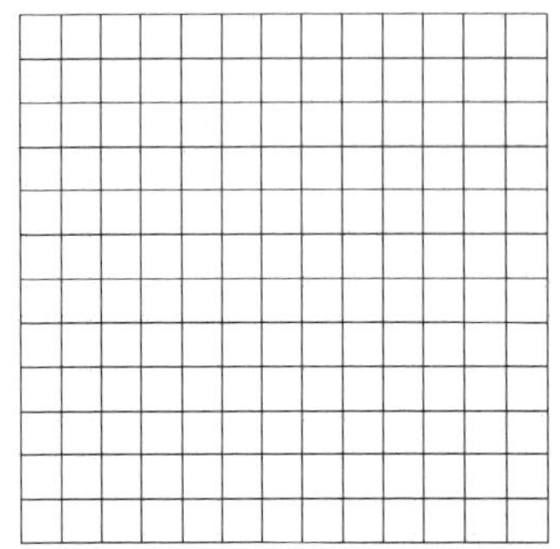

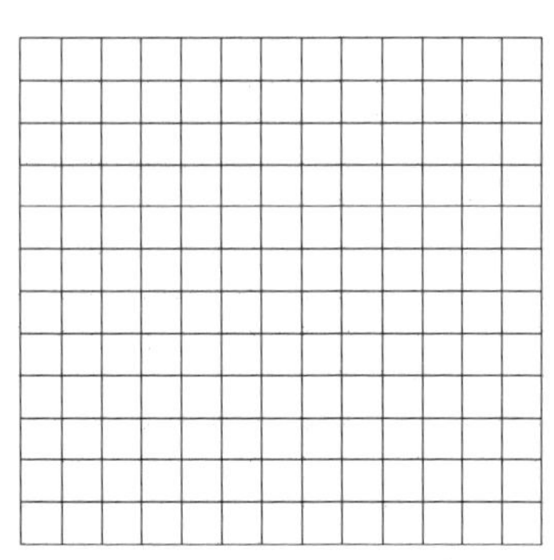

1	2	3	4	5	6	7	8
N	O	E	P	R	A	Q	L

1.C. Trigonometric functions

Theory and Examples

The 6 trigonometric functions are:

$\sin(x)\,;\cos(x)\,;\tan(x) = \frac{\sin(x)}{\cos(x)};\cot(x) = \frac{\cos(x)}{\sin(x)} = \frac{1}{\tan(x)};\sec(x) = \frac{1}{\cos(x)};\csc(x) = \frac{1}{\sin(x)}$

The trigonometric functions are *periodic* functions. This means they repeat themselves periodically.

EXAMPLE

The graph of sin(x) is represented below.

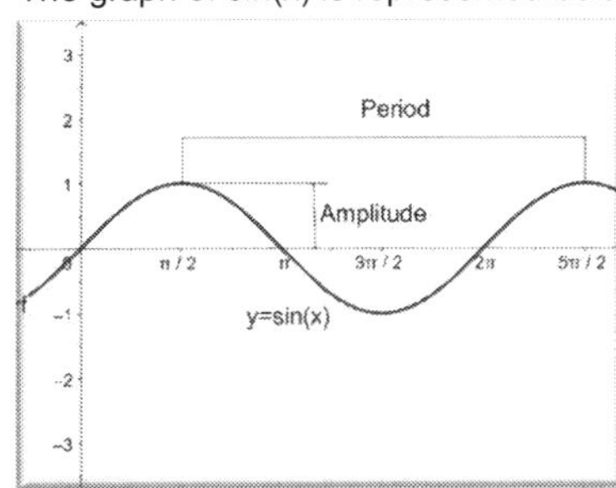

One radian is the measure of an angle subtended at the center of a circle by an arc which is equal in length to the radius of the circle as can be seen in figure below.

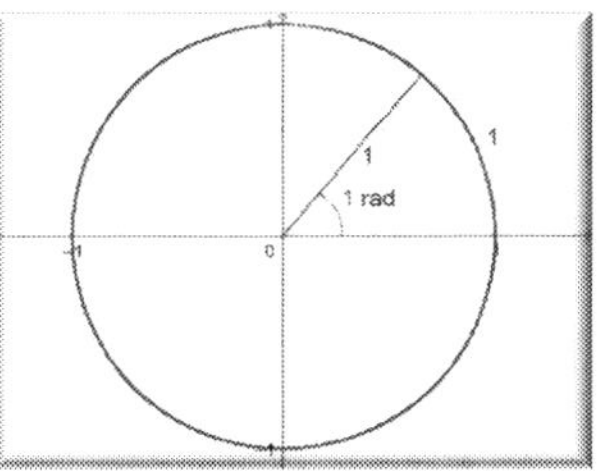

$2\pi\ radians = 360^0 so, \pi\ radians = 180^0$

As can be seen below is the period for

$f(x) = \sin(x)\ and\ g(x) = \cos(x)\ is\ 2\pi, instead\ h(x) = \tan(x)$ has a period of π.

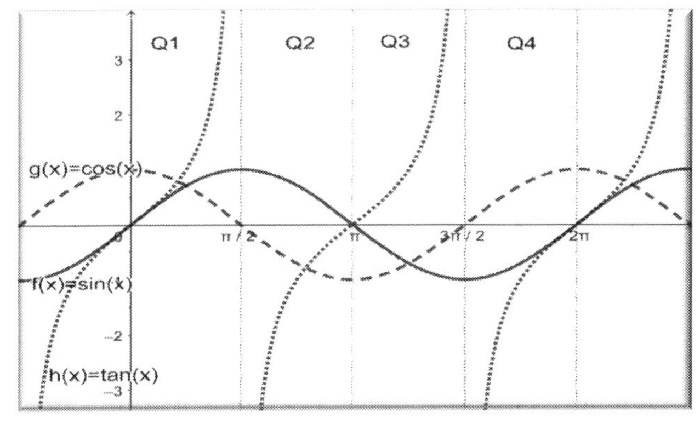

Q1 – first quadrant
Q2 – second quadrant
Q3 – third quadrant
Q4 – fourth quadrant

The sign of functions sin(x); cos(x) and tan(x) in each quadrant is shown below.

	Q1	Q2	Q3	Q4
Sin(x)	+	+	-	-
Cos(x)	+	-	-	+
Tan(x)	+	-	+	-

EXAMPLE

Tan(x) is negative in quadrants Q2 and Q4.

In a right-angle triangle, with $\sphericalangle\phi \neq 90^0$ we have the following *trigonometric ratios*:

$\sin\sphericalangle\phi = \frac{opposite}{hypotenuse}$

$\cos\sphericalangle\phi = \frac{adjacent}{hypotenuse}$

$\tan\sphericalangle\phi = \frac{opposite}{adjacent}$

$\cot\sphericalangle\phi = \frac{adjacent}{opposite}$

EXAMPLE

In the right-angle triangle $\Delta ABC\ with \sphericalangle B = 90^0$.

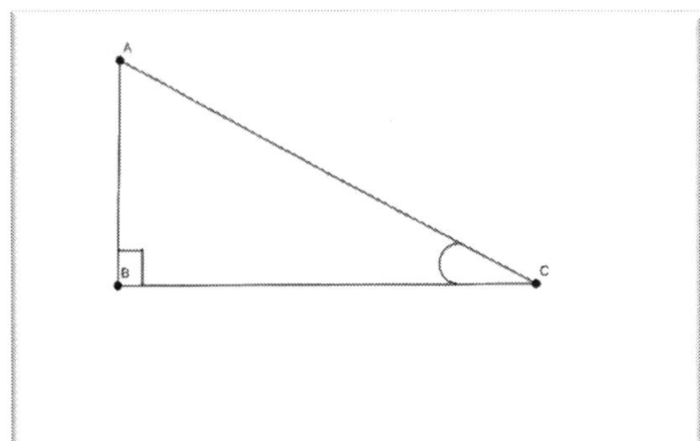

$\sin\sphericalangle C = \frac{opposite}{hypotenuse} = \frac{AB}{AC}$

$\cos\sphericalangle C = \frac{adjacent}{hypotenuse} = \frac{BC}{AC}$

$\tan\sphericalangle C = \frac{opposite}{adjacent} = \frac{AB}{BC}$

$\cot\sphericalangle\phi = \frac{adjacent}{opposite} = \frac{BC}{AB}$

Chapter 1. C. Trigonometric functions

How long are the legs of a foal compared with a mature horse? They are approximately the.......

Determine which statement is correct. In the table at the bottom of the page cross all the letters of the correct answers off.

1) The length of the part of a trigonometric function that repeats, measured along the x axis is called **period.**

2) One radian is the measure of an angle subtended at the center of a circle by an arc which is equal in length to the radius of the circle.

3) $\pi\ radians = 360^0$

4) In Quadrant 1 $\sin(\alpha)$ is positive.

5) In Quadrant 3 $\tan(\alpha)$ is negative.

6) If $\cos(\alpha_1) = k, k \geq 0$ the other value of α that is a solution of the equation is:

$\alpha_2 = 2\pi - \alpha_1$ (radians).

7) $\cos(\alpha) = \frac{opposite}{adjacent}$

8) $\tan(\alpha) = \frac{opposite}{adjacent}$

9) $\sin\left(\frac{\pi}{2}\right) = 1$

10) $\cos\left(\frac{\pi}{4}\right) = 1$

1	2	3	4	5	6	7	8	9	10
O	L	S	I	A	V	M	R	C	E

1.D. Graphs of trigonometric functions

a Graphing sine and cosine functions

Theory and Examples

Sine and cosine trigonometric functions have the domain: all real values of x, and the range: interval [-1, 1].

EXAMPLE

The graph of $f(x) = \sin(x)$ is shown below.

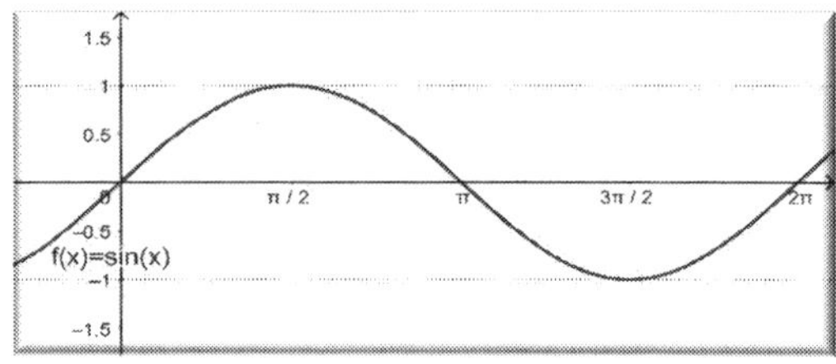

The value of $\sin\left(\frac{3\pi}{2}\right) = -1 \mp 2n\pi, n\ is\ an\ integer.$

The value of $\sin(2\pi) = 0 \mp 2n\pi, n\ is\ an\ integer.$

EXAMPLE

The graph of $f(x) = \cos\,(x)$ is shown below.

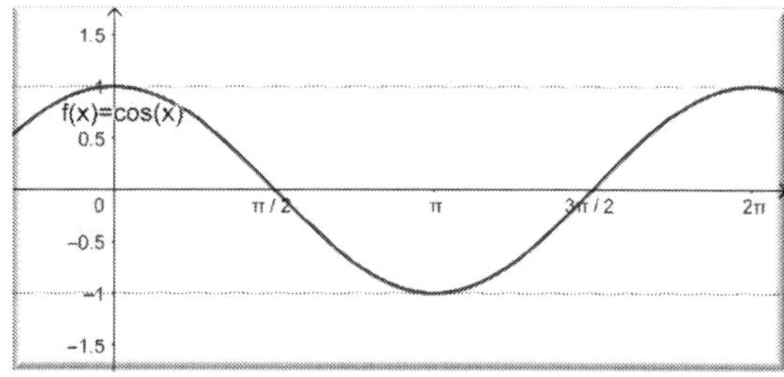

The value of $\cos\left(\frac{3\pi}{2}\right) = 0 \mp 2n\pi, n\ is\ an\ integer.$

The value of $\sin(2\pi) = 1 \mp 2n\pi, n\ is\ an\ integer.$

Chapter 1. D. a. Graphing sine and cosine functions

Who was one of the most persistent enemies of Gaius Julius Caesar whom, after a life time of opposing Caesar in the Roman senate as member of "optimates" and always Caesar's equal in persistence, and force of character, died in Africa at Utica (Tunisia) in 46 B.C. two years before Caesar was assassinated?

Determine which statement is correct. In the table at the bottom of the page cross all the letters of the correct answers off.

1) For each of values of α, the values of $\sin(\alpha)$ are:

α	0	$\frac{\pi}{6}$	$\frac{\pi}{2}$	$\frac{\pi}{3}$	π
$\sin(\alpha)$	0	$\frac{1}{2}$	1	$\frac{\sqrt{3}}{2}$	0

2) The graph of $\sin(\alpha)$ is:

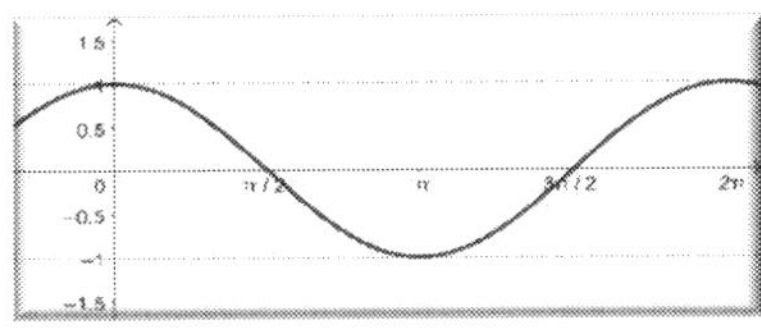

3) The minimum of $\cos(\alpha)$ is -1

4) The maximum of $\sin(\alpha)$ is -1

5) The graph of $\cos(\alpha)$ is:

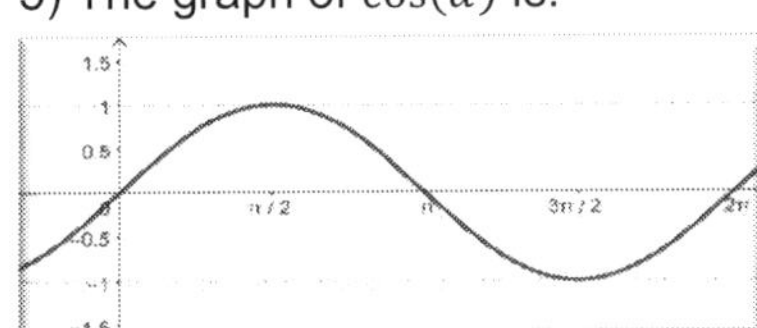

6) For each of values of α, the values of $\cos(\alpha)$ are:

α	0	$\frac{\pi}{6}$	$\frac{\pi}{2}$	$\frac{\pi}{3}$	π
$\cos(\alpha)$	1	$\frac{1}{2}$	2	$\frac{1}{2}$	0

1	2	3	4	5	6
B	C	D	A	T	O

1.D. Graphs of trigonometric functions

b Graphing tangent and cotangent functions

Theory and Examples

Tangent and cotangent trigonometric functions have the domain all the real values for x but the ones where either sine or cosine functions are zero.
For values of $x = \frac{\pi}{2} \mp k\pi, k - integer$ the tangent is undefined. For these values of x there are vertical asymptotes. The range is: all real numbers.
<u>Asymptotes</u> are lines where graphs of the functions go towards zero and the distance between the graph and asymptote approaches zero but never becomes zero.
The tangent graph is shown below.

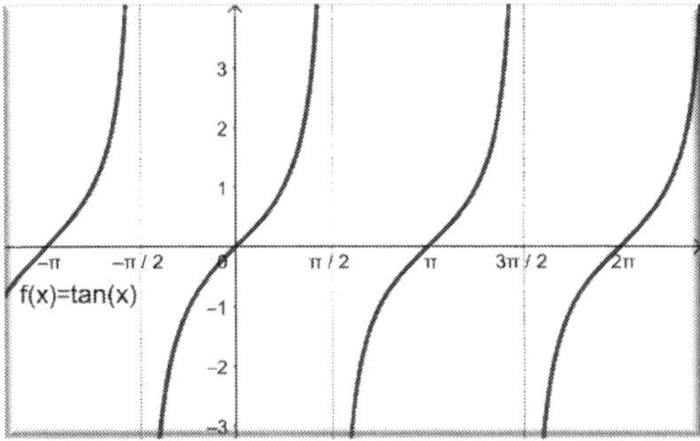

EXAMPLE

The tangent of zero is zero.
We know that $\tan(x) = \frac{\sin(x)}{\cos(x)}$ so, the tangent function is zero when the sine function is zero. The tangent function is <u>undefined</u> when the cosine function is zero.

EXAMPLE

Sine function is zero for $x = 0, and\ tangent\ is\ zero\ for\ x = \mp k\pi\ where\ k - integer.$
We know that $\cot(x) = \frac{\cos(x)}{\sin(x)}$. The cotangent function is undefined for $x = \mp k\pi, k - integer.$
The sign functions Sin(Φ), Cos(Φ), and Tan(Φ) in each quadrant is shown below.

	Q1	Q2	Q3	Q4
Sin(x)	+	+	-	-
Cos(x)	+	-	-	+
Tan(x)	+	-	+	-

The values for the special angles in a right-angle triangle are given below.

	30^0	60^0	45^0
Sin(Φ)	$\frac{1}{2}$	$\frac{\sqrt{3}}{2}$	$\frac{\sqrt{2}}{2}$
Cos(Φ)	$\frac{\sqrt{3}}{2}$	$\frac{1}{2}$	$\frac{\sqrt{2}}{2}$
Tan(Φ)	$\frac{\sqrt{3}}{3}$	$\sqrt{3}$	1

EXAMPLE

If $\sin(30^0) = \frac{1}{2}$ and, $\cos(30^0) = \frac{\sqrt{3}}{2}$ then:

$$\tan(30^0) = \frac{\sin(30^0)}{\cos(30^0)} = \frac{\frac{1}{2}}{\frac{\sqrt{3}}{2}} = \frac{1}{2} \div \frac{\sqrt{3}}{2} = \frac{1}{2} \times \frac{2}{\sqrt{3}} = \frac{1}{\sqrt{3}} = \frac{\sqrt{3}}{3}$$

The cotangent graph is shown below.

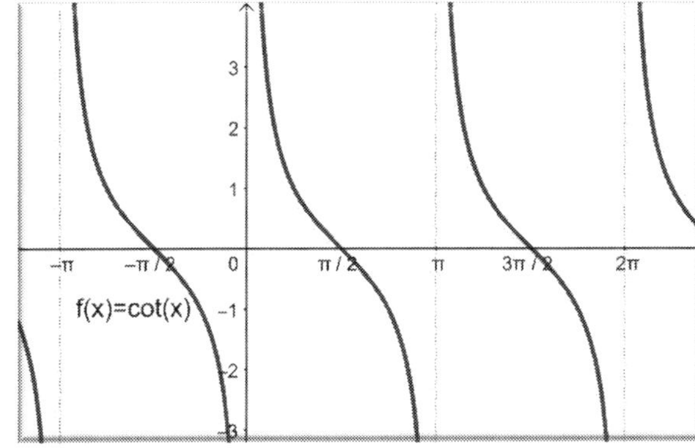

For values of $x = \pi \mp k\pi, k - integer$ the cotangent is undefined. For these values of x, there are vertical asymptotes.

EXAMPLE

If $\sin(\mathbf{30^0}) = \frac{1}{2}$ and, $\cos(\mathbf{30^0}) = \frac{\sqrt{3}}{2}$ then:

$$\cot(\mathbf{30^0}) = \frac{\cos(30^0)}{\sin(30^0)} = \frac{\frac{\sqrt{3}}{2}}{\frac{1}{2}} = \frac{\sqrt{3}}{2} \div \frac{1}{2} = \frac{\sqrt{3}}{2} \times \frac{2}{1} = \sqrt{3}$$

If $\sin(\mathbf{60^0}) = \frac{\sqrt{3}}{2}$ and, $\cos(\mathbf{60^0}) = \frac{1}{2}$ then:

$$\cot(\mathbf{60^0}) = \frac{\cos(\mathbf{60^0})}{\sin(\mathbf{60^0})} = \frac{\frac{1}{2}}{\frac{\sqrt{3}}{2}} = \frac{1}{2} \div \frac{\sqrt{3}}{2} = \frac{1}{2} \times \frac{2}{\sqrt{3}} = \frac{1}{\sqrt{3}} = \frac{\sqrt{3}}{3}$$

Chapter 1. D. b. Graphing tangent and cotangent functions

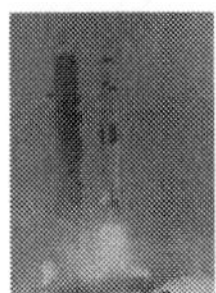

What was the rocket that lifted the Apollo capsule called?

Determine which statement is correct. In the table at the bottom of the page cross all the letters of the correct answers off.

1) The tangent function is undefined at angles of 180^0 and 270^0

2) The tangent graph looks like the one below for $-\frac{\pi}{2} < \propto < \frac{\pi}{2}$

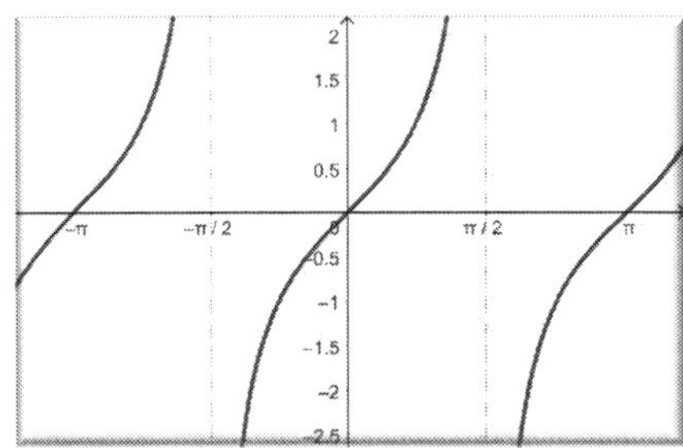

3) The cotangent function is zero for $\frac{\pi}{2} + k\pi, k\ integer.$

4) The tangent function is zero for $\frac{\pi}{2} + k\pi, k\ integer.$

5) For 60^0 the tangent is $\sqrt{3}$

6) Tangent of 45^0 is 1

7) Cotangent of 30^0 is 2

8) The tangent function has an amplitude of 100

9) Cotangent of 45^0 is 1

10) The formula of tangent in terms of sine and cosine is $\tan(\propto) = \frac{\cos(\propto)}{\sin(\propto)}$

11) Tangent of 30^0 is $\frac{\sqrt{3}}{3}$

12) Cotangent of $60^0$is $\frac{1}{2}$

1	2	3	4	5	6
S	B	C	A	D	Q
7	8	9	10	11	12
T	U	X	R	E	N

1.E. Inverse trigonometric functions

Theory and Examples

Remember that an *inverse* function $f^{-1}(x)$ is the function that has the Domain equal to the Range of the original function $f(x)$. The Range of the inverse function equals the Domain of the original function.

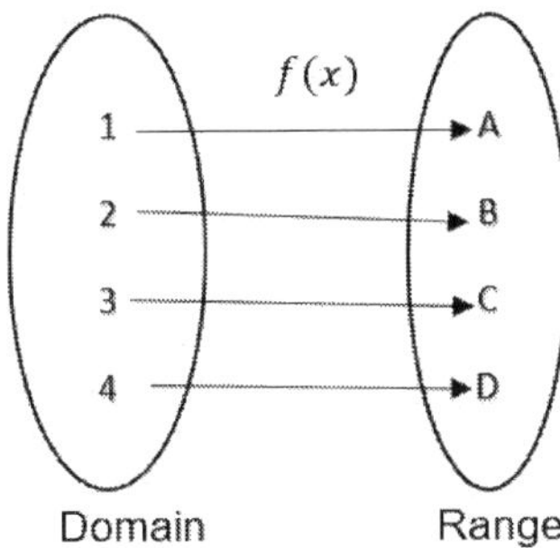

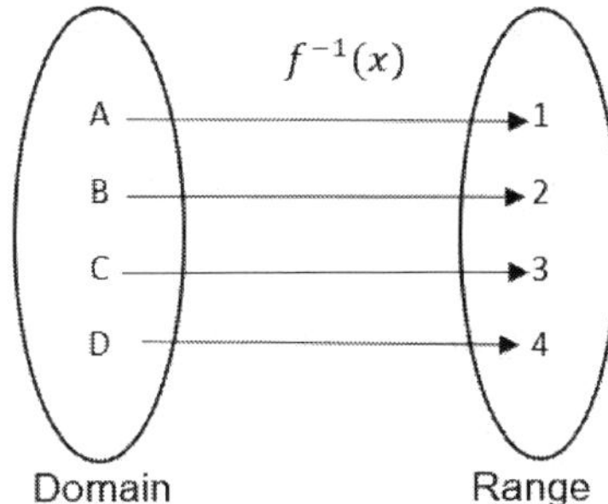

EXAMPLE

Because we are talking about trigonometric functions like sine, cosine and tangent, in the notation for inverse trigonometric function,

we substitute $f^{-1}(x)$ with $\Phi = sin^{-1}(\mathrm{x})$, where:

x is value of the original trigonometric function.

Φ is the angle we are looking for.

EXAMPLE

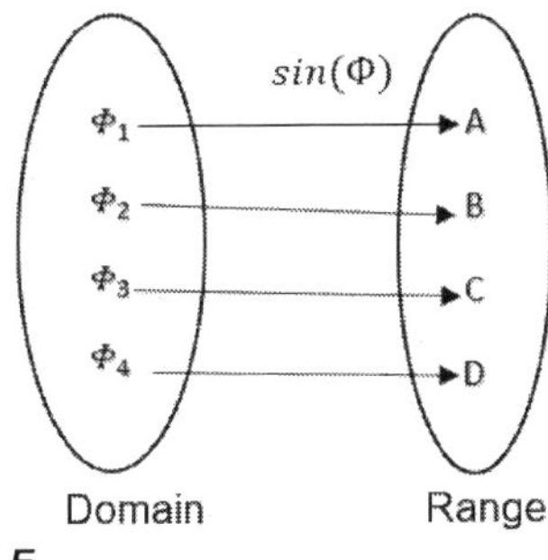

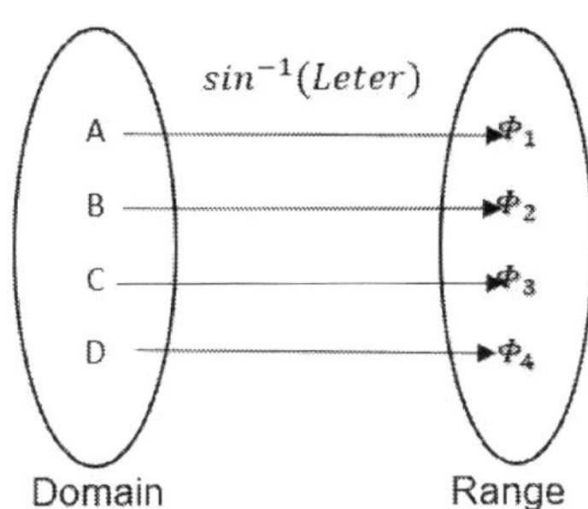

EXAMPLE

If $\sin(\Phi) = 0.345$, then $\Phi = sin^{-1}(0.345) = 20.18^0$

If $\tan(\Phi) = 1.89$, then $\Phi = tan^{-1}(1.89) = 62.11^0$

Chapter 1. E. Inverse Trigonometric Functions

What is the name of the mountain that is the tallest in the Canadian Rockies, is situated near Jasper and has height of 3,954 m?

Determine which statement is correct. In the table at the bottom of the page cross all the letters of the correct answers off.

1) Inverse trig functions do the opposite of the "regular" trig functions.

2) The angle $\propto$ in right angle triangle PSQ is: 45^0

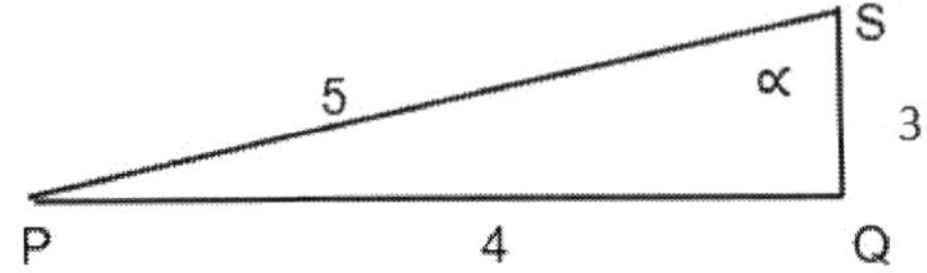

3) The angle $\propto$ in right angle triangle PSQ is: 35^0

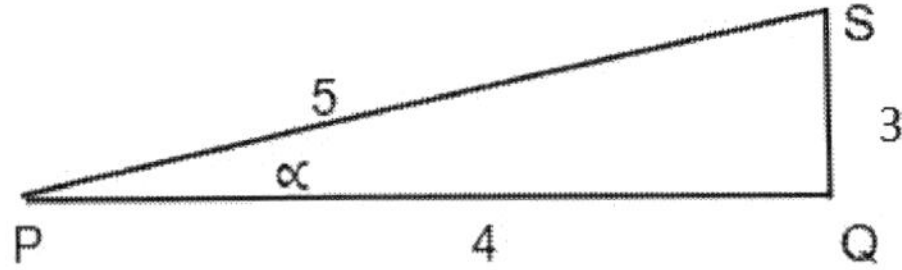

4) $Tan^{-1}(\sqrt{3})$ is 60^0

5) The domain of $f(x) = sin^{-1}(x)$ is $[0,2\pi]$

6) The range (principal value) of $f(x) = cos^{-1}(x)$ is all real numbers.

7) The domain of $f(x) = tan^{-1}(x)$ is $[-\pi,2\pi]$

8) The angle $\propto$ in the figure bellow is 20^0

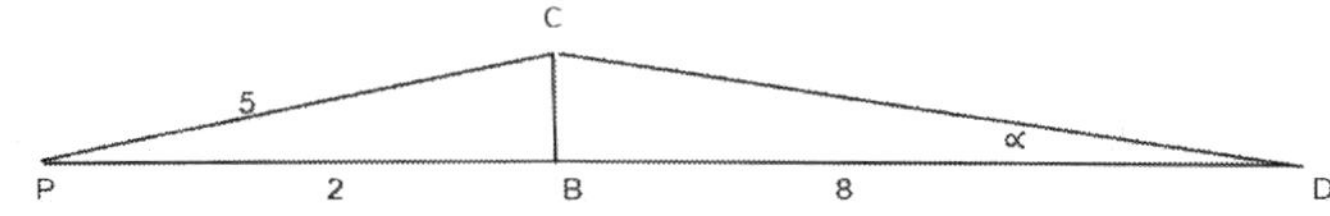

9) The $cos^{-1}\left(\frac{\sqrt{2}}{2}\right)$ is 45^0

10) The $sin^{-1}(0.3)$ is 65^0

1	2	3	4	5	6	7	8	9	10
T	R	O	C	B	E	S	O	A	N

1.F. Graphs of inverse trigonometric functions

Theory and Examples

The inverse function of sine is $f^{-1}(x) = sin^{-1}(x) = \arcsin(x)$.
The graph of $f^{-1}(x) = sin^{-1}(x) = \arcsin(x)$ is shown below.

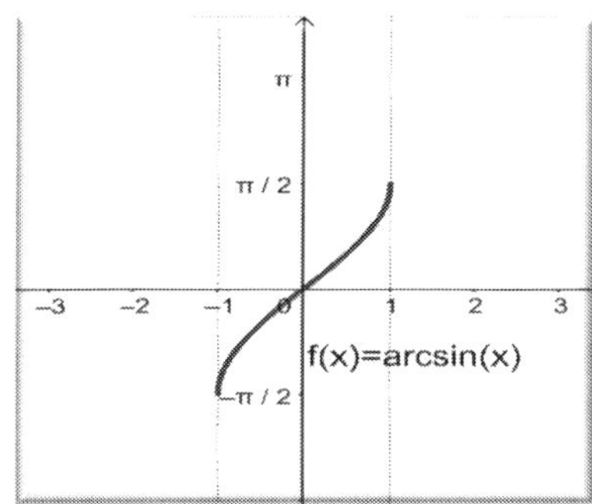

The domain is the interval [-1, 1].
The range is the interval $\left[-\frac{\pi}{2}, \frac{\pi}{2}\right]$.

EXAMPLE

$f(0) = \arcsin(0) = 0$
$f(-1) = \arcsin(-1) = -\frac{\pi}{2}$

The graph of $f^{-1}(x) = cos^{-1}(x) = \arccos(x)$ is shown below.

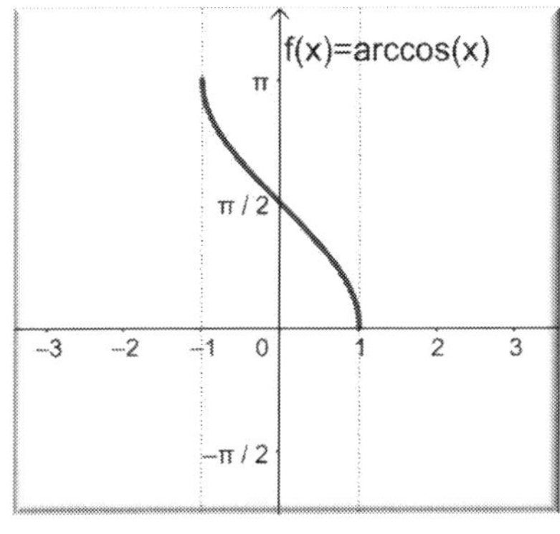

The domain is the interval [-1, 1].
The range is the interval $[0, \pi]$.

EXAMPLE

$f(-1) = \arccos(-1) = \pi$

The graph of $f^{-1}(x) = tan^{-1}(x) = \arctan(x)$ is shown below.

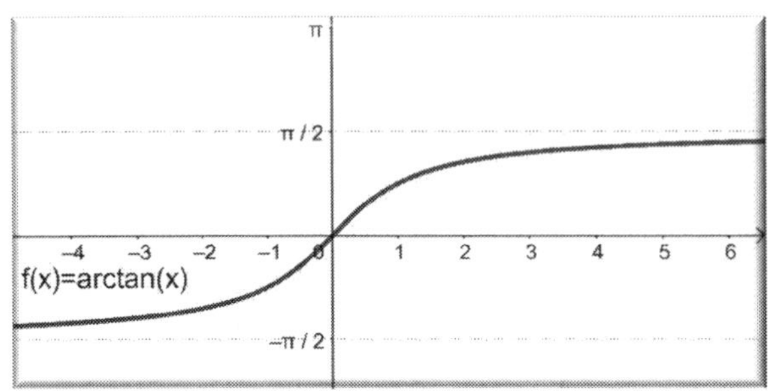

The domain is the real numbers.

The range is the interval $\left[-\frac{\pi}{2}, \frac{\pi}{2}\right]$.

EXAMPLE

$f(1) = \arctan(1) = \frac{\pi}{4}$

$f(-4) = \arctan(-4) = -0.47\,\pi$

The graph of $f^{-1}(x) = cot^{-1}(x) = \text{arccot}(x)$ is shown below.

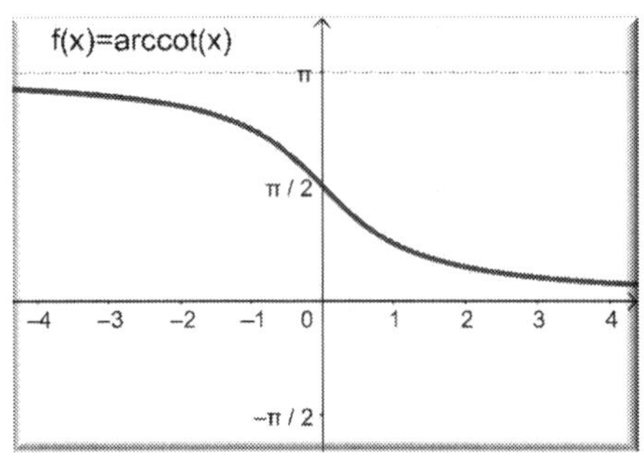

The domain is the real numbers.

The range is the interval $[0, \pi]$.

EXAMPLE

$f(-1) = \text{arccot}(-1) = 0.75\pi$

$f(-4) = \text{arccot}(-4) = 0.92\,\pi$

We could calculate the arccot(x) as $\frac{\pi}{2} - [\arctan(x)]$.

EXAMPLE

$f(-1) = \text{arccot}(-1) = \frac{\pi}{2} - [\arctan(x)] = \frac{\pi}{2} - [\arctan(-1)] = \frac{\pi}{2} - \left[-\frac{\pi}{4}\right] = \frac{\pi}{2} + \frac{\pi}{4} = \frac{3\pi}{4} = 0.75\pi$

$f(-4) = \text{arccot}(-4) = \frac{\pi}{2} - [\arctan(x)] = \frac{\pi}{2} - [\arctan(-4)] = \frac{\pi}{2} - [-0.42\pi] = 0.92\,\pi$

Chapter 1. F. Graphs of Inverse Trigonometric Functions

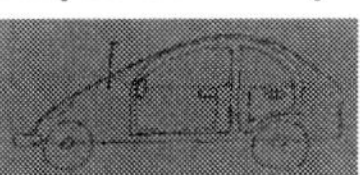

What country in the world has the most cars?

Determine which statement is correct. In the table at the bottom of the page cross all the letters of the correct answers off.

1) The domain of $f(x) = cos^{-1}(x)$ is [-1,1]

2) Graph the function $f(x) = cos^{-1}(x)$

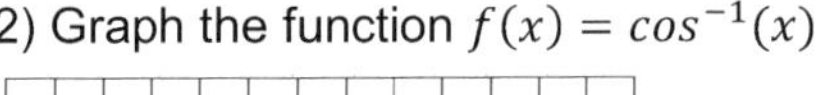

3) Graph the function $f(x) = sin^{-1}(x)$

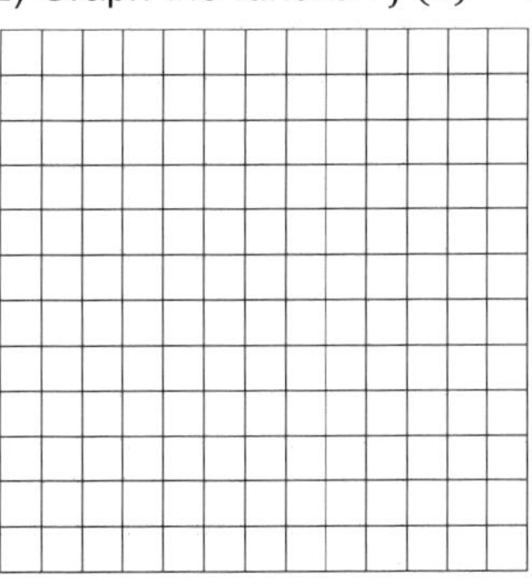

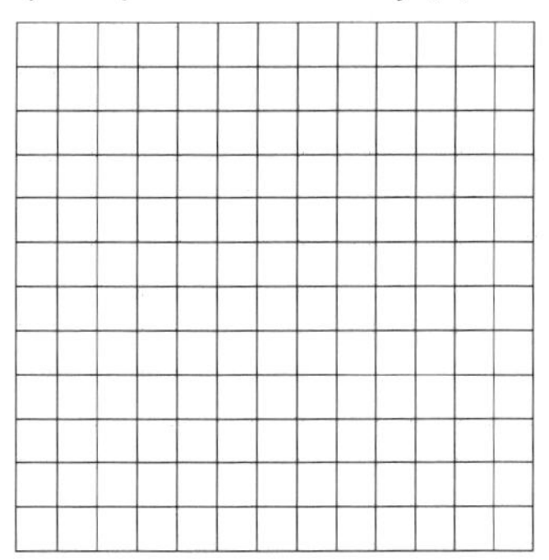

4) The value of $f^{-1}(1) = cos^{-1}(1)$ is 2

5) The graph of $f(x) = cos^{-1}(x)$ looks like the one below.

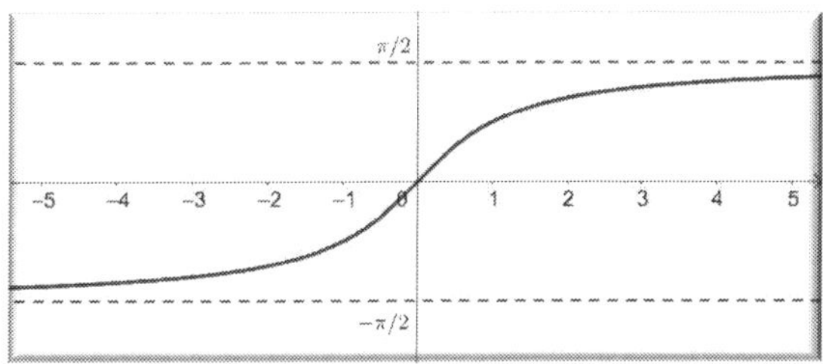

6) The value of $f(0) = cos^{-1}(0)$ is 1

1	2	3	4	5	6
O	C	R	U	S	A

QUICK ANSWERS

Chapter 1

1.A.a	DSKY
1.A.b	POMPEY
1.A.c	FORD
1.B	NEPAL
1.C	SAME
1.D.a	CATO
1.D.b	SATURN
1.E	ROBSON
1.F	USA

FULL SOLUTIONS

CHAPTER 1

Chapter 1. A. a. Determine: if a relation is a function, the values of a function, the range

1. Incorrect

For each value of the domain there is only one value that belongs to the range. For domain value of 3, there are two values that belong to the range: 3, and 12. (3,3) and (3,12).

2. Correct

For each value of the domain there is only one value that belongs to the range.

3. Incorrect

For each value of the domain there is only one value that belongs to the range. For domain value of -4, there are two values that belong to the range: 8, and 1. (-4,8) and (-4,1)

4. Correct

For each value of the domain there is only one value that belongs to the range.

5. Incorrect

For x=2 ; $f(2) = 5(2) + 3 = 10 + 3 = 13$

6. Correct

11For x=1 ; $f(1) = 3(1^2) - 3(1) + 2 = 3(1) - 3 + 2 = 3 - 3 + 2 = 2$

7. Correct

For s=2 ; $G(2) = \frac{2^2-3(2)+7}{2+3} = \frac{4-6+7}{5} = \frac{5}{5} = 1$

8. Correct

For t=-1; $G(-1) = 3 - 2(-1) = 3 + 2 = 5$
For t=-2 ; $G(-2) = 3 - 2(-2) = 3 + 4 = 7$
For t=3 ; $G(3) = 3 - 2(3) = 3 - 6 = -3$

9. Incorrect

For x=-2 ; $F(-2) = (-2)^2 - 5(-2) + 1 = 4 + 10 + 1 = 15$
For x=0 ; $F(0) = (0)^2 - 5(0) + 1 = 0 - 0 + 1 = 1$
For x=3 ; $F(3) = (3)^2 - 5(3) + 1 = 9 - 15 + 1 = 7\ not - 4$

10. Correct

For c=-3 ; $H(-3) = \frac{(-3)^2-2(-3)}{-3+2} = \frac{9+6}{-1} = -15$
For c=0 ; $H(0) = \frac{(0)^2-2(0)}{0+2} = \frac{0}{2} = 0$
For c=3 ; $H(3) = \frac{(3)^2-2(3)}{3+2} = \frac{9-6}{5} = \frac{3}{5}$

Chapter 1. A. b. Linear and quadratic functions and their graphs

1. Incorrect

The difference between each consecutive value belonging to the domain should be always the same. The difference between x=9 and x=7 is two not one as it should be. The difference between each consecutive value belonging to the range should be always the same. The difference between y=7 and y=5 is two, the difference between y=8 and y=7 is one, the difference between y=10 and y=8 is two again.

2. Correct

The difference between each consecutive value belonging to the domain and range are be always the same, not with each other.

3. Incorrect

The difference between each consecutive value belonging to the domain should be always the same. The difference between x=2 and x=2 is zero not one as it should be.

4. Correct

$f(x) = 2x + 3$ represents the expression of a linear function with slope of 2 and y intercept at y=3

5. Incorrect

The degree of the function is 2 not one. This expression represents a quadratic function.

6. Correct

$f(x) = 34x + 15$ represents the expression of a linear function with slope of 34 and y intercept at y=15

7. Correct

The graph is a straight line. It represents a linear function.

8. Incorrect

The graph is not a straight line. It represents a quadratic function.

9. Incorrect

The graph is not a straight line.

10. Correct

The expression represents a quadratic function.

11. Correct

The expression represents a quadratic function.

12. Incorrect

The expression doesn't represent a quadratic function. The degree of the function is 4 not 2 as it should be.

Chapter 1. A. c. Inverse functions and their graphs

1. Correct

$y = 2x + 3$ so, we interchange x with y and have:
$x = 2y + 3$; minus 3 each side
$x - 3 = 2y$; divide with 2 both sides
$\frac{x-3}{2} = y\,; so\; f^{-1}(x) = \frac{x-3}{2}$

2. Incorrect

$y = 3x - 5$ so, we interchange x with y and have:
$x = 3y - 5$; + 5 each side
$x + 5 = 3y$; divide with 3 both sides
$\frac{x+5}{3} = y\,; so\; f^{-1}(x) = \frac{x+5}{3}$

3. Incorrect

$y = x^2 - 1$, so, we interchange x with y and have
$x = y^2 - 1$; plus 1 each side
$x + 1 = y^2$; square root of both sides
$\sqrt{x+1} = y\colon x \geq -1; so\; f^{-1}(x) = \sqrt{x+1}$

4. Correct

$y = \frac{3}{2x+4}$; so, we interchange x with y and have

$x = \frac{3}{2y+4}$; cross multiply and have

$x(2y + 4) = 3$; use the distributivity property and have

$2xy + 4x = 3$; subtract 4x from both sides

$2xy = 3 - 4x$; divide both sides with 2x

$y = \frac{3 - 4x}{2x}$ $so\ f^{-1}(x) = \frac{3 - 4x}{2x}$

5. Incorrect

$y = \frac{5}{2x^2+4}$; so, we interchange x with y and have

$x = \frac{5}{2y^2+4}$; cross multiply and have

$x(2y^2 + 4) = 5$; use the distributivity property and have

$2xy^2 + 4x = 5$; subtract 4x from both sides

$2xy^2 = 5 - 4x$; divide both sides with 2x

$y^2 = \frac{5-4x}{2x}$; square root both sides

$y = \mp\sqrt{\frac{5-4x}{2x}}; so\ f^{-1}(x) = \mp\sqrt{\frac{5-4x}{2x}}; x > 0, x \leq 5/4$

6. incorrect

$y = \frac{\sqrt{x-1}}{3}$; so, we interchange x with y and have

$x = \frac{\sqrt{y-1}}{3}$; multiply with 3 both sides

$3x = \sqrt{y-1}$ square both sides

$9x^2 = y - 1$; add 1 in both sides

$9x^2 + 1 = y\ so\ f^{-1}(x) = 9x^2 + 1$

7. Correct

$y = \frac{3x}{x+3}$; so, we interchange x with y and have

$x = \frac{3y}{y+3}$; cross multiply and have

$x(y + 3) = 3y$; use the distributivity property and have

$xy + 3x = 3y$; subtract xy in each side

$3x = 3y - xy$ Take y as common factor

$3x = y(3 - x)$; divide with (3-x) in both sides

$\frac{3x}{3 - x} = y\ so\ f^{-1}(x) = \frac{3x}{3 - x}$

8. Correct

$y = \frac{1}{5x+3}$; so, we interchange x with y and have

$x = \frac{1}{5y+3}$; cross multiply and have

$x(5y+3) = 1$; use the distributivity property and have

$5xy + 3x = 1$; subtract 3x in both sides

$5xy = 1 - 3x$; divide by 5x both sides

$y = \frac{1-3x}{5x}$; $so\ f^{-1}(x) = \frac{1-3x}{5x}$,

Graph the following functions and their inverse functions on the same graph.

9. $f(x) = y = \frac{2}{3x+2}$ the inverse is: $f^{-1}(x) = \frac{2-2x}{3x}$

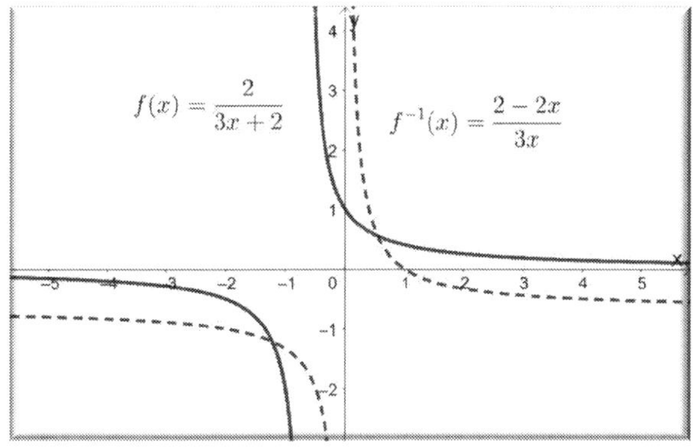

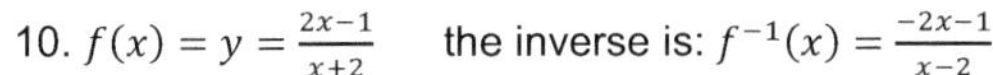

10. $f(x) = y = \frac{2x-1}{x+2}$ the inverse is: $f^{-1}(x) = \frac{-2x-1}{x-2}$

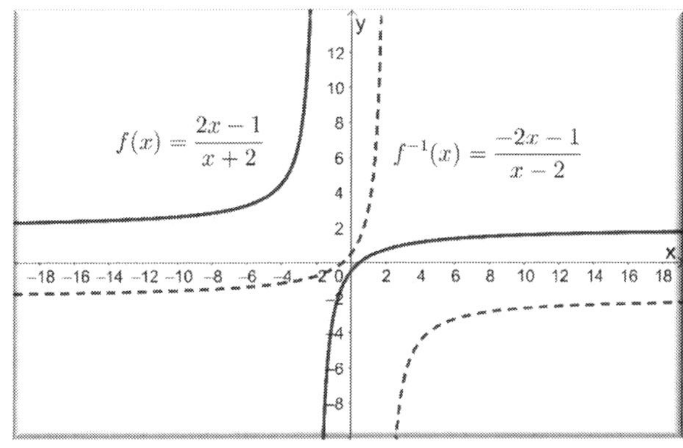

Chapter 1. B. Piecewise functions

1. Incorrect

$f(x) = 2x^2 + 5x - 3\,, x \in R$ is a parabola.

2. Correct

$f(x) = \begin{cases} 2x \; for \; x < 0 \\ 3 \; for \; x \geq 0 \end{cases}$ is defined by multiple sub-functions. Each of these sub-functions is spread over to a certain interval or part of the function's domain.

3. Incorrect

$f(x) = 4x + 3, x \in R$ is a straight line.

4. Incorrect

$f(x) = 7x - 3, x \in R$ is a straight line.

5. Correct

$f(x) = \begin{cases} 2x^4 \; for \; x < -2 \\ 3x + 6 \; for \; x \geq -2 \end{cases}$ is defined by multiple sub-functions. Each of these sub-functions is spread over to a certain interval or part of the function's domain.

6. Incorrect

$f(x) = 3\sin(x - 2) + 3, x \in R$ is a periodic continuous function.

7. Correct

$f(x) = \begin{cases} x^2 \; for \; x \leq -2 \\ 4 \; for -2 < x < 2 \\ 2x \; for \; x \geq 2 \end{cases}$ is defined by multiple sub-functions. Each of these sub-functions is spread over to a certain interval or part of the function's domain.

8. Incorrect

$f(x) = \sqrt{x - 1}$ is a continuous function for x>1

Graph function 5 and 7.

<u>Function 5</u>

$$f(x) = \begin{cases} 075x^4 \; for \; x < 2 \\ 3x + 6 \; for \; x \geq 2 \end{cases}$$

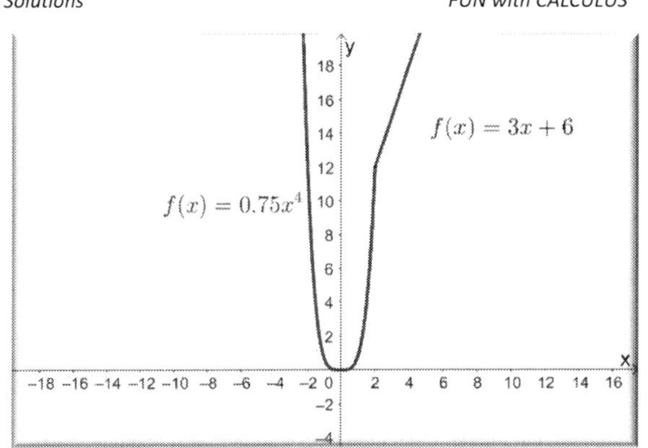

Function 7

$$f(x) = \begin{cases} x^2 \, for \, x \leq -2 \\ 4 \, for \, -2 < x < 2 \\ 2x \, for \, x \geq 2 \end{cases}$$

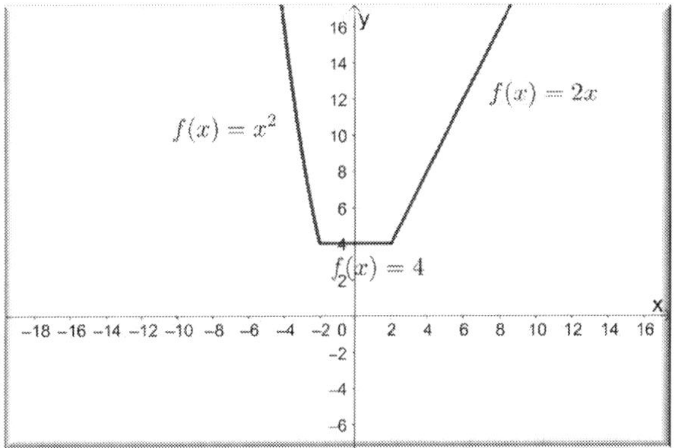

Chapter 1. C. Trigonometric functions

1. Correct

The length of the part of a trigonometric function that repeats, measured along the x axis is called **period.**

2. Correct

One radian is the measure of an angle subtended at the center of a circle by an arc which is equal in length to the radius of the circle.

3. Incorrect

$\pi\ radians = 180^0$

4. Correct

In Quadrant 1 when α is between zero and 90 degrees or $\frac{\pi}{2}$, $\sin(\alpha)$ is positive. See below.

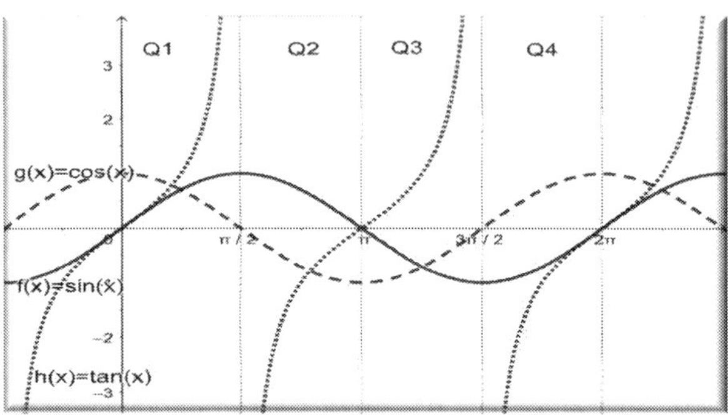

5. Incorrect

In Quadrant 3 where when α is between 180^0 or $\pi\ rad$ and 270^0 or $\frac{3\pi}{2}\ rad$, $\tan(\alpha)$ is positive. See below.

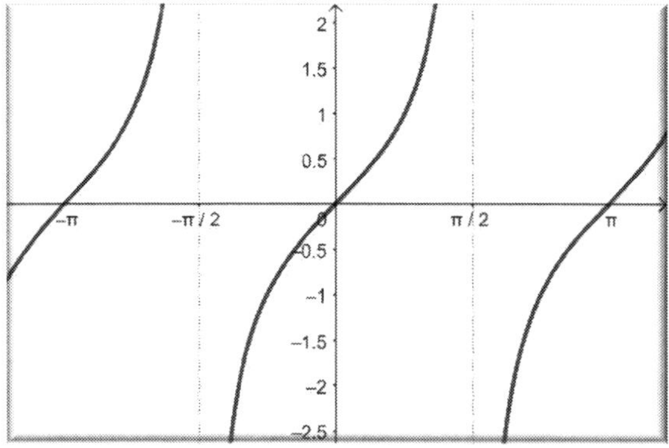

6. Correct

If $\cos(\alpha_1) = k, k \geq 0$ the other value of α that is solution of the equation is:$\alpha_2 = 2\pi - \alpha_1$ (radians). See below.

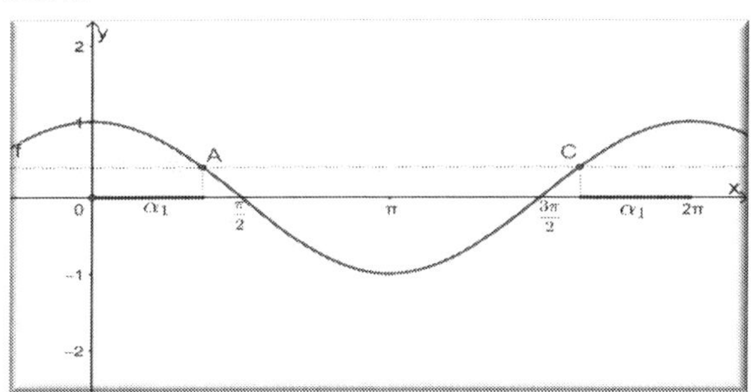

7. Incorrect

In a right-angle triangle $\cos(\alpha) = \frac{adjacent}{hypotenuse}$

8. Correct.

In a right-angle triangle $\tan(\alpha) = \frac{opposite}{adjacent}$

9. Correct

$\sin\left(\frac{\pi}{2}\right) = 1.$

10. Incorrect.

$\cos\left(\frac{\pi}{4}\right) = \frac{\sqrt{2}}{2} = \frac{1.41}{2} = 0.707\ not\ 1$

The values for the special angles in a right-angle triangle are given below.

	30^0	60^0	45^0
Sin(Φ)	$\frac{1}{2}$	$\frac{\sqrt{3}}{2}$	$\frac{\sqrt{2}}{2}$
Cos(Φ)	$\frac{\sqrt{3}}{2}$	$\frac{1}{2}$	$\frac{\sqrt{2}}{2}$
Tan(Φ)	$\frac{\sqrt{3}}{3}$	$\sqrt{3}$	1

Chapter 1. D. a. Graphing sine and cosine functions

1. Correct

For each of values of α, the values of $\sin(\alpha)$ are:

α	0	$\frac{\pi}{6} = 30^0$	$\frac{\pi}{2} = 90^0$	$\frac{\pi}{3} = 60^0$	π
$\sin(\alpha)$	0	$\frac{1}{2}$	1	$\frac{\sqrt{3}}{2}$	0

2. Incorrect

The graph of $\sin(\alpha)$ is:

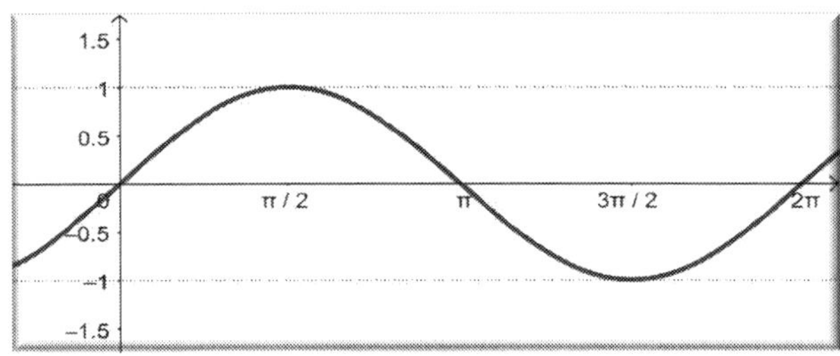

3. Correct

The minimum of $\cos(\alpha)$ is -1

4. Incorrect

The maximum of $\sin(\alpha)$ is +1

5. Incorrect

The graph of $\cos(\alpha)$ is:

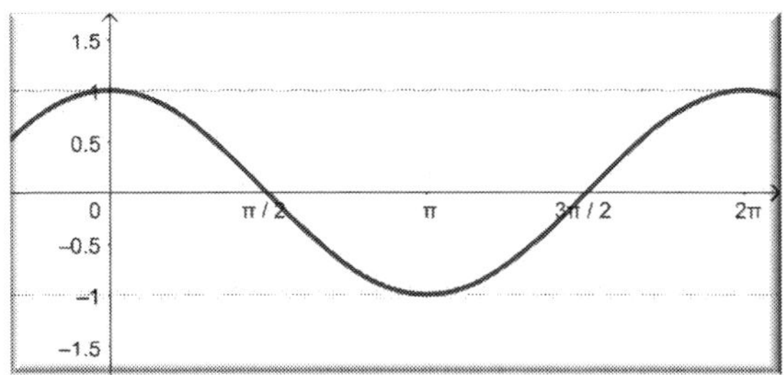

6. Incorrect

For each of values of α, the values of $\cos(\alpha)$ are:

α	0	$\frac{\pi}{6} = 30^0$	$\frac{\pi}{2}$	$\frac{\pi}{3} = 60^0$	π
$\cos(\alpha)$	1	$\frac{\sqrt{3}}{2}$	0	$\frac{1}{2}$	-1

Chapter 1. D. b. Graphing tangent and cotangent functions

1. Incorrect

The tangent function is defined for angle $180^0\ (= 0)$ and $undefined\ for\ 270^0$

2. Correct

The tangent graph looks like the one below for $-\frac{\pi}{2} < \alpha < \frac{\pi}{2}$

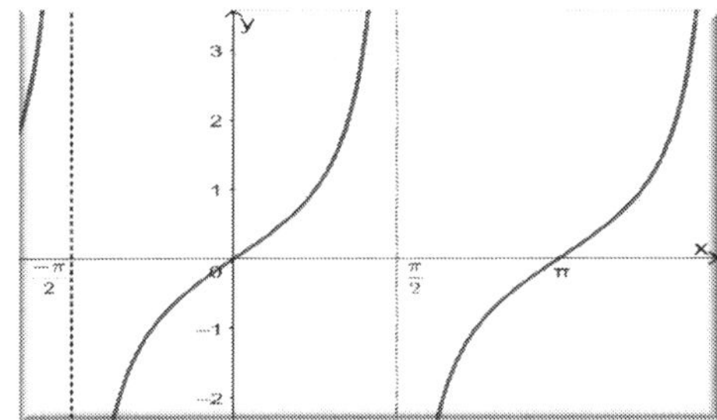

3. Correct

The cotangent function is zero for

$\frac{\pi}{2} + k\pi, k\ integer$

4. Incorrect

The tangent function is undefined for $\frac{\pi}{2} + k\pi, k\ integer.$

5. Correct

For 60^0 the tangent is $\sqrt{3}$

6. Correct

Tangent of 45^0 is 1

7. Incorrect

Cotangent of 30^0 is $\sqrt{3}$

8. Incorrect

The tangent function does not have an amplitude.

9. Correct

Cotangent of 45^0 is 1

10. Incorrect

The formula of tangent in terms of sine and cosine is $\tan(\propto) = \frac{\sin(\propto)}{\cos(\propto)}$

11. Correct

Tangent of 30^0 is $\frac{\sqrt{3}}{3}$

12. Incorrect

Cotangent of 60^0 is $\frac{\sqrt{3}}{3}$

Chapter 1. E. Inverse Trigonometric Functions

1. Correct

Inverse trig functions do the opposite of the "regular" trig functions.

2. Incorrect

The angle $\propto$ in right angle triangle PSQ is: 53.2^0

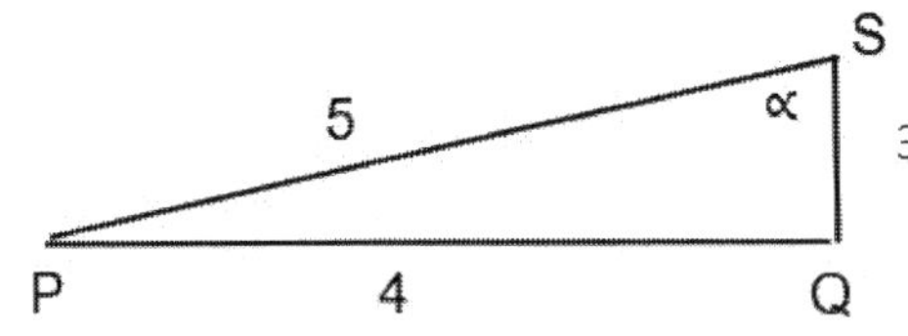

$sin \propto = \frac{4}{5} = 0.8\,; so\ \alpha = sin^{-1}(0.8) = 53.2^0$

3. Incorrect

The angle $\propto$ in right angle triangle PSQ is: 36.8^0

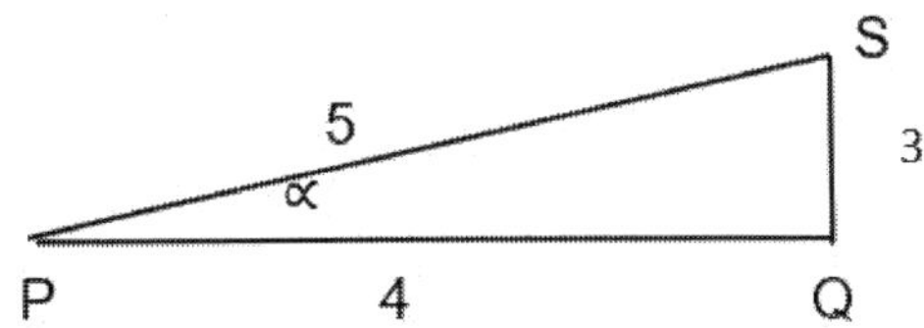

$cos \propto = \frac{4}{5} = 0.8\,; so\ \alpha = cos^{-1}(0.8) = 36.8^0$

4. Correct

$Tan^{-1}(\sqrt{3})$ is 60^0

5. Incorrect

The domain of $f(x) = sin^{-1}(x)$ is [-1,1]

6. Correct

The range (principal value) of $f(x) = cos^{-1}(x)$ is all real numbers.

7. Incorrect

The domain of $f(x) = tan^{-1}(x)$ is all real numbers.

8. Incorrect

The angle ∝ in the figure bellow is 29.7^0

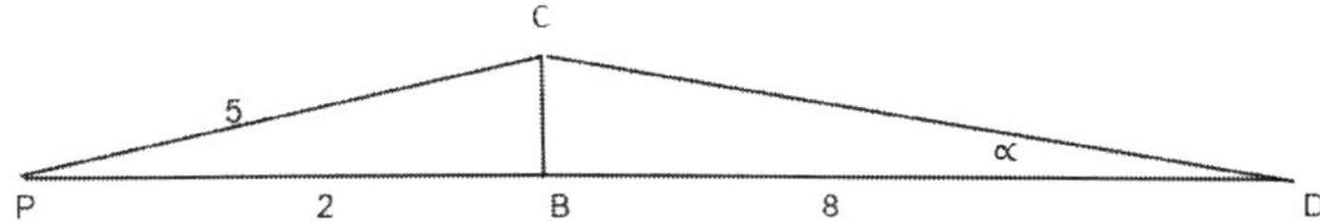

Step 1

In the right-angle triangle PBC we are using the Pythagorean theorem. So, we have:

$CB^2 = CP^2 - PB^2 = 5^2 - 2^2 = 25 - 4 = 21\,; so\ CB = \sqrt{21} = 4.58$

Step 2

In the right-angle triangle CBD $tan \propto = \frac{4.58}{8} = 0.57\,; so\ tan^{-1}(0.57) = 29.7^0$

9. Correct

The $cos^{-1}\left(\frac{\sqrt{2}}{2}\right)$ is 45^0

10. Incorrect

The $sin^{-1}(0.3)$ is 17.45^0

Chapter 1. F. Graphs of Inverse Trigonometric Functions

1. Correct

The domain of $f(x) = cos^{-1}(x)$ is [-1,1]

2. Graph the function $f(x) = cos^{-1}(x)$

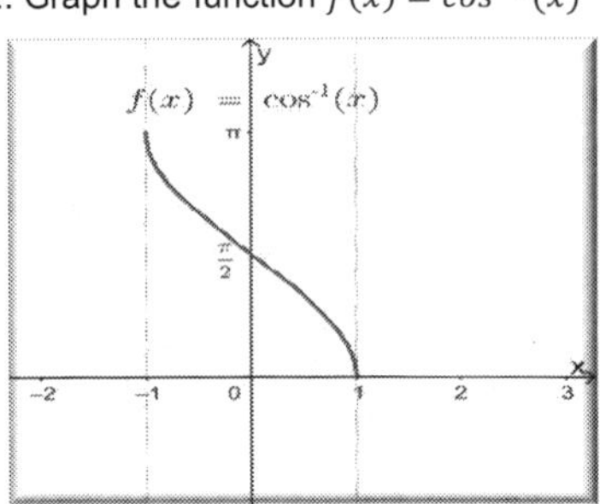

3. Graph the function $f(x) = sin^{-1}(x)$

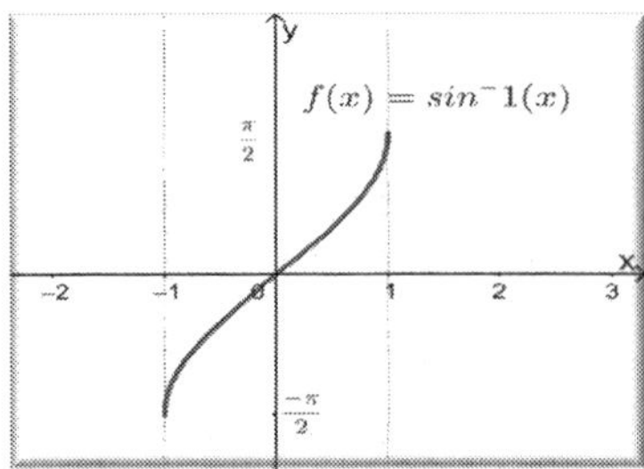

4. Incorrect

The value of $f(1) = cos^{-1}(1)$ is 0

5. Incorrect

The graph of $f(x) = cos^{-1}(x)$ looks like the one in problem 2 above.

6. Incorrect

The value of $f(0) = cos^{-1}(0)$ is $\frac{\pi}{2}$

CHAPTER 2

Limits

"Cogito ergo sum." (latin)

(I think, therefore I am.)

Rene Descartes (1596 - 1650)

(Invented the Cartesian system of axes)

Questions to be answered:

- What is one of the biggest differences between the black bear and the grizzly bear? What does the grizzly have that the black bear doesn't?
- He was one of the men that Caesar liked and helped a lot, yet he was the Roman senator out of twenty-two senators that killed Caesar with the last blow. Who was this infamous senator?
- The Sea of Tranquility is located on the Moon. The sea doesn't have water, and it is a lower-altitude plain. What is another name for "sea" that is used in the Apollo Program?
- This German inventor built the first modern automobile in 1886. What is his family name?
- On which continent, other than North America, do elks live?
- Which is the smallest of the bear species that has black fur?
- What is the name of the volcano that is the tallest in Japan and has a height of 3776.24 m?
- In 83 BC the Dictator of Rome asked young Julius Caesar to divorce his beloved wife Cornelia. Young Caesar disobeyed this formidable and dangerous Dictator of Rome and left Rome in exile until the Dictator's death in 78 BC. Who was this Dictator of Rome?

Fundamental Concepts

UNDERSTANDING LIMITS:

To understand limits better let's analyze this example. We are watching a soccer game. There is a penalty kick. The player is approaching the ball. The TV transmission is interrupted for 2 seconds. When the transmission is back, the soccer player already scored. The ball is rolling towards the center of the post. What happened while the transmission was off? What did the soccer player do to score? What path did the ball take? Using limits, we can approximate the path the ball followed. Let's split the time interval into very small parts on both images, before and after the player scores. We then compare the path of the ball before and after the transmission interruption. In this way, the path the ball took can be approximated easier.

Now,

If we have a function f(x) that gets very close to a value L for a value of x that gets very close to a value "a", we call L the limit of function f for x getting very close to a certain value "a".

EXAMPLE

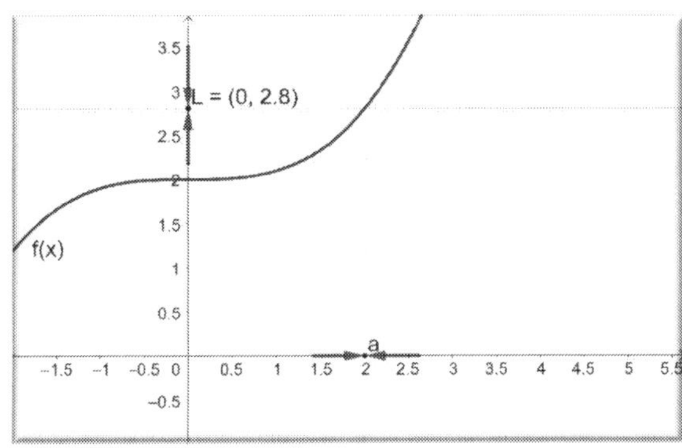

The function f(x) is getting very close to the value L for x approaching very close to a value "a".

Point L (0,2.8) is the limit of function f(x) when x is going very close to value a=2.

EXAMPLE

Suppose $f(x) = 0.1x^3 + 2$

We want to see what happens when x is very close to value 2. We create a table of values in such a way that we take values for x closer and closer to value 2, on each side of 2, smaller and bigger than 2.

The table below shows the values for x around 2 and for f(x) around the limit L=2.8.

As we can see, as we approach x=2 from both sides f(x) is approaching L=2.8

X	$f(x) = 0.1x^3 + 2$
1.5	2.3375
1.9	2.6859
1.99	2.7881
1.999	2.7988
2.001	2.8012
2.01	2.8121
2.1	2.9261

A geometric illustration

Let's suppose we have a circle. The chord AB is the longest chord compared with the others in this example. As we are approaching point A going around the circle from point B towards point A, through points B, C, D, E, the length of the chord is becoming smaller and smaller. When we are at a point which is extremely close to point A, the length of the chord is extremely small. The moment we are in point A the chord becomes a tangent to the circle, line AT.

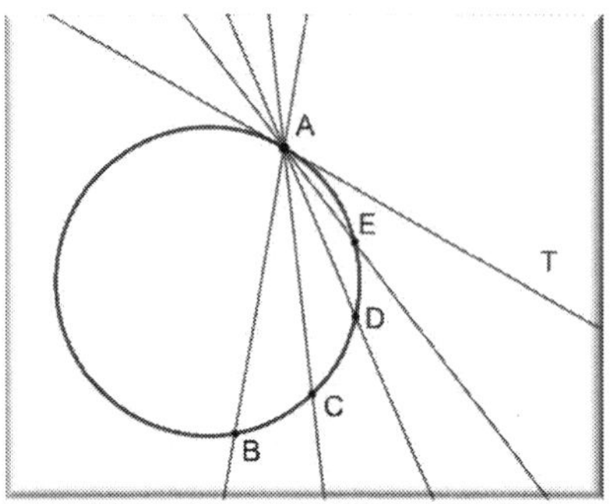

The word tangent comes from the Latin word **tangens** which means "touching". Any tangents to any curve, touch that particular curve in one and *only* one point.

Chapter 2. A. a. Limits – Table of values, and graphically

What is one of the biggest differences between the black bear and the grizzly bear? What does the grizzly have that the black bear doesn't?

Determine if the statements below are correct. Cross the letters off for the correct answers.

1) We can show limits through graphs or tables of values.

2) If x=2 doesn't belong to the domain, the function doesn't have limit for f(2).

3) The table below represent the value of $f(x) = 0.3x^2 - 2$ around -0.8 for x around 2

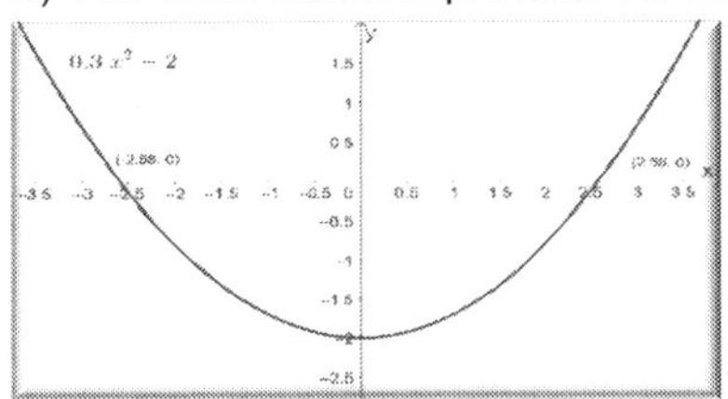

X	$f(x) = 0.3x^2 - 2$
1.8	-1.028
1.9	-0.917
1.999	-0.8012
2.001	-0.7988
2.01	-0.7879
2.1	-0.677

4) The table below represents the value of $f(x) = x^2 + 3$ around 7.6 for x around 5

X	$f(x) = x^2 + 3$
4.8	27
4.99	27.5
4.999	27.6
5.001	27.9
5.01	28
5.1	29

5) The table above represent the value of $f(x) = x^2 + 3$ around 25 for x around 5

6) The table below represents the value of $f(x) = x^2 + 2x$ around 10 for x around 2

X	$f(x) = x^2 + 3$
1.89	6.5
1.999	6.59
2.001	6.61
2.01	6.62
2.1	6.9

1	2	3	4	5	6
A	B	R	U	M	P

2.A. Limits

b. Algebraically

Theory and Examples

The symbol used when we calculate the limit L of a function f(x) for a value of x approaching a value "a" is:

$\lim_{x \to a} f(x) = L$

Simplistically, when the function has values (Range) for each value of real values of x (Domain), that function is called continuous.

When x equals a certain value where the function exists and is continuous, we just substitute that value of x in the expression of the limit.

EXAMPLE

$$\lim_{x \to 2} \frac{x^2+2x+4}{x+3} = \lim_{x \to 2} \left[\frac{(x+2)^2}{x+3}\right] = \frac{\lim_{x \to 2}(x+2)^2}{\lim_{x \to 2}(x+3)} = \frac{(2+2)^2}{(2+3)} = \frac{16}{5} = 3\frac{1}{5}$$

There are **three cases** for when we have to find the limits of a rational expression algebraically for x approaches plus or minus infinity.

1. When the degree of the polynomial at the numerator **is less** than the degree of the polynomial at the denominator. In this case, the limit will always be zero for x going towards ∓∞. The graph will have a horizontal asymptote x=0.

EXAMPLE

Find $\lim_{x \to \infty} \frac{x+1}{x^2-2}$

$$\lim_{x \to \infty} \frac{x+1}{x^2-2} = \lim_{x \to \infty} \frac{x+1}{x^2-2} = \lim_{x \to \infty} \left[\frac{x^2(\frac{1}{x}+\frac{1}{x^2})}{x^2(1-\frac{2}{x^2})}\right] = \frac{\lim_{x \to \infty}\left(\frac{1}{x}\right)+\lim_{x \to \infty}\left(\frac{1}{x^2}\right)}{\lim_{x \to \infty} 1-\lim_{x \to \infty}\left(\frac{2}{x^2}\right)} = \frac{0+0}{1-0} = \frac{0}{1} = 0$$

2. When the degree of the polynomial at the numerator **is the same** as the degree of the polynomial at the denominator. In this case, the limit will always equal the ratio between the coefficients of the terms with the highest exponent for x going towards ∓∞. The graph will have a horizontal asymptote at x= the ratio between the coefficients of the terms with the highest exponent.

EXAMPLE

Find $\lim_{x \to \infty} \frac{3x^2+1}{x^2-2}$

$$\lim_{x \to \infty} \frac{3x^2+1}{x^2-2} = \lim_{x \to \infty} \frac{3x^2+1}{x^2-2} = \lim_{x \to \infty} \left[\frac{x^2(3+\frac{1}{x^2})}{x^2(1-\frac{2}{x^2})}\right] = \frac{\lim_{x \to \infty}(3)+\lim_{x \to \infty}\left(\frac{1}{x^2}\right)}{\lim_{x \to \infty} 1-\lim_{x \to \infty}\left(\frac{2}{x^2}\right)} = \frac{3+0}{1-0} = \frac{3}{1} = 3$$

3. When the degree of the polynomial at the numerator **is greater** than the degree of the polynomial at the denominator. In this case, the graph will tend to go either upward or downward, depending on the signs of the terms with the highest exponents from the numerator and denominator.

EXAMPLE

Find $\lim\limits_{x\to\infty}\frac{3x^3+1}{x^2-2}$

$$\lim_{x\to\infty}\frac{3x^3+1}{x^2-2}=\lim_{x\to\infty}\frac{3x^3+1}{x^2-2}=\lim_{x\to\infty}\left[\frac{x^2(3x+\frac{1}{x^2})}{x^2(1-\frac{2}{x^2})}\right]=\frac{\lim\limits_{x\to\infty}(3x)+\lim\limits_{x\to\infty}(\frac{1}{x^2})}{\lim\limits_{x\to\infty}1-\lim\limits_{x\to\infty}(\frac{2}{x^2})}=\frac{\infty+0}{1-0}=\infty$$

As we can see, both terms $3x^3$ and x^2 are positive. In this case, the result will be positive so, the graph will go toward plus infinity.

EXAMPLE

$$\lim_{x\to\infty}\left(\sqrt{3x^2-2}-\sqrt{7x}\right)=\lim_{x\to\infty}\left(\sqrt{3x^2-2}-\sqrt{7x}\right)\left(\frac{\sqrt{3x^2-2}+\sqrt{7x}}{\sqrt{3x^2-2}+\sqrt{7x}}\right)=\lim_{x\to\infty}\left(\frac{3x^2-2-7x}{\sqrt{3x^2-2}+\sqrt{7x}}\right)=\lim_{x\to\infty}\frac{3x^2-7x-2}{\sqrt{3x^2+1}+\sqrt{7x}}=$$

$$\lim_{x\to\infty}\frac{x(3x-7-\frac{2}{x})}{x(\sqrt{3+\frac{1}{x^2}}+\sqrt{\frac{7}{x}})}=\lim_{x\to\infty}\frac{3x-7-\frac{2}{x}}{\sqrt{3+\frac{1}{x^2}}+\sqrt{\frac{7}{x}}}=\frac{\lim\limits_{x\to\infty}(3x-7-\frac{2}{x})}{\lim\limits_{x\to\infty}(\sqrt{3+\frac{1}{x^2}}+\sqrt{\frac{7}{x}})}=\frac{\infty-7-0}{\sqrt{3+0}+0}=\infty$$

EXAMPLE

Find $\lim\limits_{h\to0}\frac{f(x+h)-f(x)}{h}$ $when\ f(x)=x^2-2x$

$$\lim_{h\to0}\frac{f(x+h)-f(x)}{h}=\lim_{h\to0}\frac{[(x+h)^2-2(x+h)]-(x^2-2x)}{h}=\lim_{h\to0}\frac{x^2+2hx+h^2-2x-2h-x^2+2x}{h}\lim_{h\to0}\frac{2hx+h^2-2h}{h}=\lim_{h\to0}\frac{h(2x-h-2)}{h}=$$

$$\lim_{h\to0}(2x-h-2)=2x-2$$

Chapter 2. A. b. Limits – Algebraically

He was one of the men that Caesar liked and helped a lot, yet he was the Roman senator, of twenty-two senators, that killed Caesar with the last blow. Who was this infamous senator?

Determine which limit is correct. In the table at the bottom of the page cross off all the letters for the correct answers.

1) $\lim_{x\to 0}(\frac{6x^2-3}{x+3}) = 3$

2) $\lim_{x\to 3}(\frac{9x^2-1}{3x-1}) = 10$

3) $\lim_{x\to 2}\frac{x^2+2x-8}{x-2} = 7$

4) $\lim_{x\to 3}\left(\frac{1}{x-3} - \frac{6}{x^2-9}\right) = \frac{1}{2}$

5) $\lim_{x\to\infty}\frac{4x+3}{x^2+4x-2} = 0$

6) $\lim_{x\to\infty}\left(\sqrt{2x+1} - \sqrt{2x}\right) = 2$

7) $\lim_{x\to 3}(\frac{16x^2-1}{4x-1}) = 13$

8) $\lim_{x\to 1}\left(\frac{x^4-1}{x-1}\right) = 4$

9) $\lim_{x\to 1}\left(\frac{x^4-1}{x^2-1}\right) = 5$

10) $\lim_{h\to 0}\frac{f(x+h)-f(x)}{h}$ $when\ f(x) = 3x^2 + 4x\ is\ 2x + 4$

1	2	3	4	5
B	A	R	U	K
6	7	8	9	10
T	O	D	U	S

2.B. One side versus two sides

Theory and Examples

As we saw before, when we talk about limits towards a certain value “a”, we have to take into consideration that the values x approaching a certain value “a” could be bigger or smaller compared with that value ”a”. As we can see in the graph to the left.

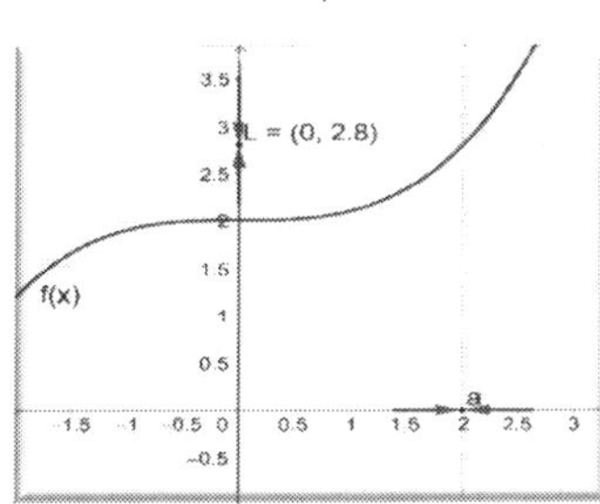

There are cases when we could have the function defined only for values bigger than the value where we want to know the limit.

EXAMPLE

The radical function.

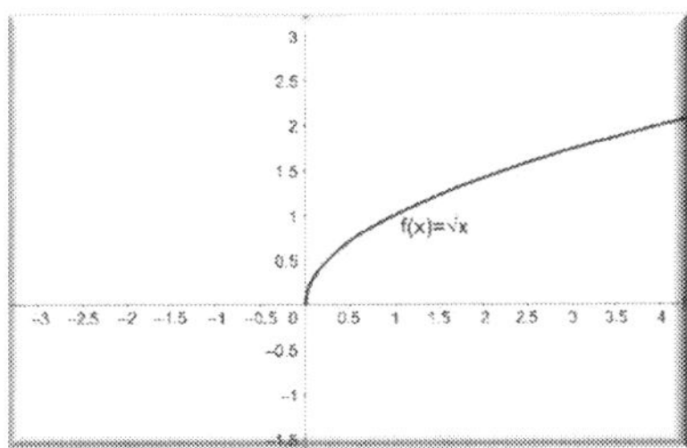

Here the values of x are bigger that zero. The limit exists only for x bigger than zero.

We call this limit **one side limit.**

The notation for this limit is: $\lim_{x \to 0^+} f(x)$

There are cases when we could have the function defined for values larger and smaller than the value where we want to know the limit, but the limits are different.

EXAMPLE

In this case the left limit of f(x) for values of x<1 is 1

$\lim_{x \to 1^-} f(x) = 1$

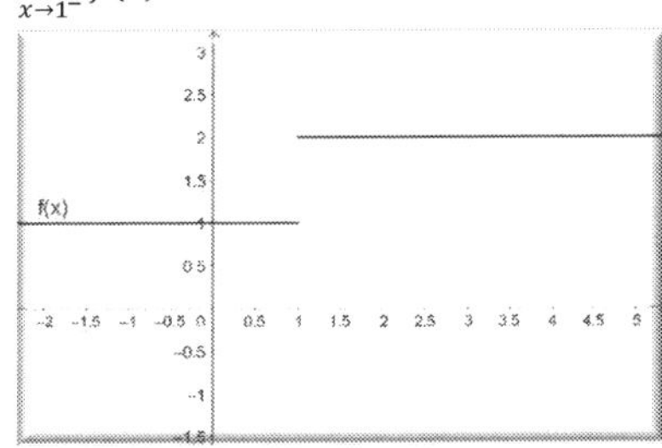

The right limit of f(x) for values of x>1 is 2

$\lim_{x \to 1^+} f(x) = 2$

There is no limit for x=1

As we can see,

$\lim_{x \to 1^-} f(x) \neq \lim_{x \to 1^+} f(x)$

Chapter 2. B. Limits-One side versus two sides

The Sea of Tranquility is located on the Moon. The sea doesn't have water, and it is a lower-altitude plain. What is another name for "sea" that is used in Apollo Program?

Determine which answer is correct. In the table at the bottom of the page cross off all the letters for the correct answers.

In the graph below $g(x) = -(0.5x - 3)^2 + 3\ for\ x < 5$

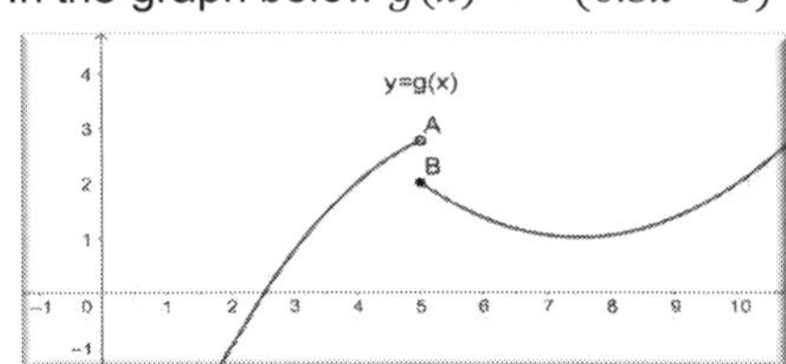

$g(x) = (0.4x - 3)^2 + 1\ for\ x \geq 5$

1) $\lim_{x\to 5^-} g(x) = 2.75$

2) $\lim_{x\to 5^+} g(x) = 2$

3) $\lim_{x\to 5} g(x) = 5$

In the graph below

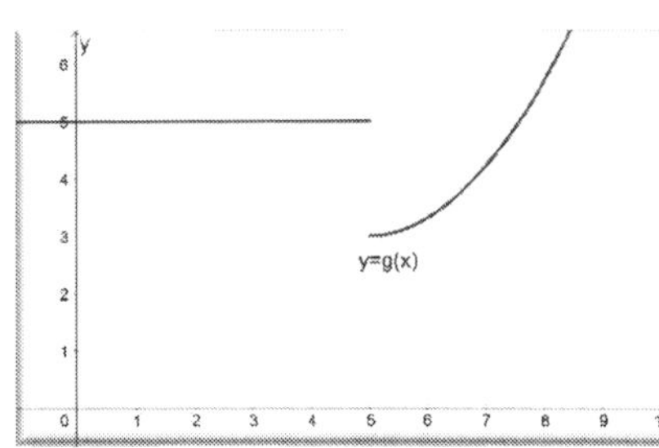

4) $\lim_{x\to 5^-} g(x) = 5$

5) $\lim_{x\to 5^+} g(x) = 3$

6) $\lim_{x\to 5} g(x) = 5$

7) $\lim_{x\to 5^+} \frac{5x}{x-5} = +\infty$

8) $\lim_{x\to 5^-} \frac{5x}{x-4} = 0$

9) $\lim_{x\to 3^-} \frac{5x}{x-5} = -7.5$

10) $\lim_{x\to 3^-} \frac{x}{x-5} = 7.5$

1	2	3	4	5	6	7	8	9	10
B	U	M	O	B	A	C	R	I	E

2.C. End behavior

Theory and Examples

This section of the chapter is all about what happens with the graph of the function when x is approaching plus or minus infinity. The question will be: what is the limit of the function when x approaches $\pm\infty$?

As x goes to $\pm\infty$, the limit of the function could be infinite, finite or does not exit.

In case the function is polynomial, there are a few scenarios.

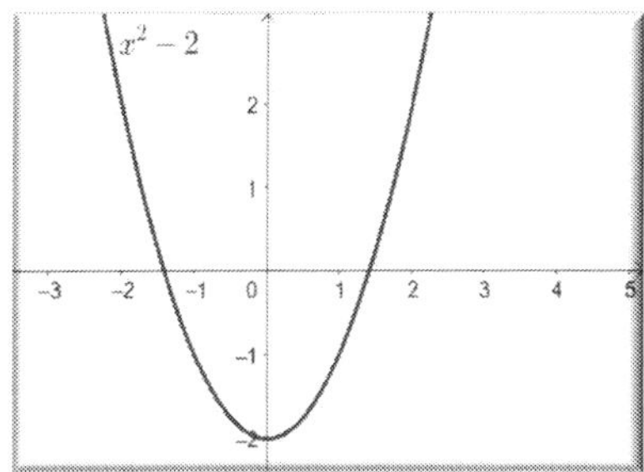

a) The degree of the function is even and the coefficient of the term with the highest degree is positive.
The graph goes upward in quadrant I and II.

b) The degree of the function is even and the coefficient of the term with the highest degree is negative.
The graph goes downward in quadrant III and IV.

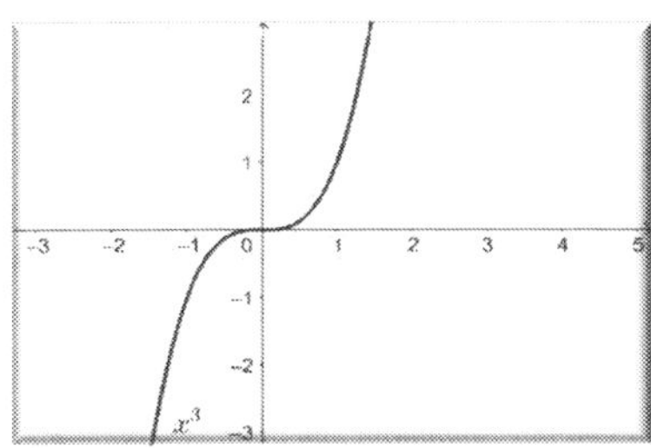

c) The degree of the function is odd and the coefficient of the term with the highest degree is positive, the graph goes upward in quadrant I and downward in quadrant III.

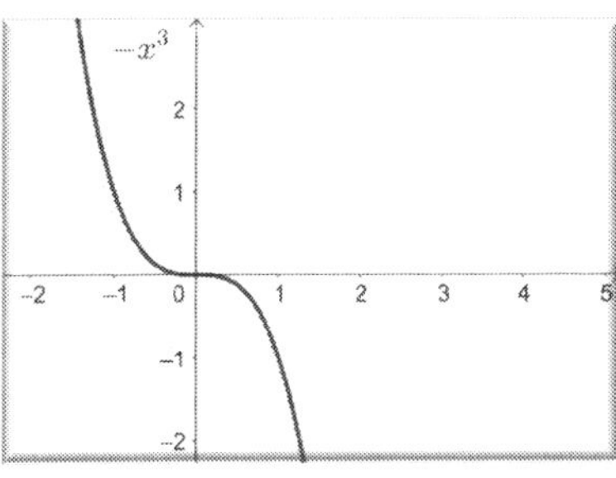

d) The degree of the function is odd and the coefficient of the term with the highest degree is negative, the graph goes upward in quadrant II and downward in quadrant IV.

Chapter 2. C. Limits-End behavior

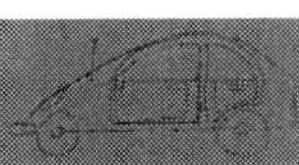

This German inventor built the first modern automobile in 1886. What is his family name?

Determine which answer is correct. In the table at the bottom of the page cross off all the letters for the correct answers.

Determine what is the end behavior of these functions using limits.

1) $f(x) = 2x^4 - 3x^2 + 15x - 24$: when x goes to +∞ it goes up in first quadrant

2) $f(x) = -4x^4 + 2x^2 - 5x$: when x goes to -∞ it goes down in the third quadrant

3) $f(x) = -4x^3 + 6x^2 - x$: when x goes to -∞ it goes up in the first quadrant

4) $f(x) = x^3 - 36x^2 - x + 73$: when x goes to -∞ it goes down in the third quadrant

5) $f(x) = \frac{2x^3+3x^2-4x+5}{6x^3-5x+4}$: when x goes to +∞ it goes towards 1

6) $f(x) = \frac{x^3+2x^2-3x+4}{6x^4-5x^3+4x^2-3x}$: when x goes to -∞ it goes towards 1

7) $f(x) = \frac{x^5+2x^2-3x+4}{6x^4-5x^3+4x^2-3x}$: when x goes to +∞ it goes up in the first quadrant

8) $f(x) = \frac{2x^3+3x^2-4x+5}{6x^2-5x+4}$: : when x goes to -∞ it goes down in the first quadrant

Graph the following functions and determine the end behavior for x going towards +∞

9) $f(x) = \frac{x^3-2x^2+3x}{x^4+1}$

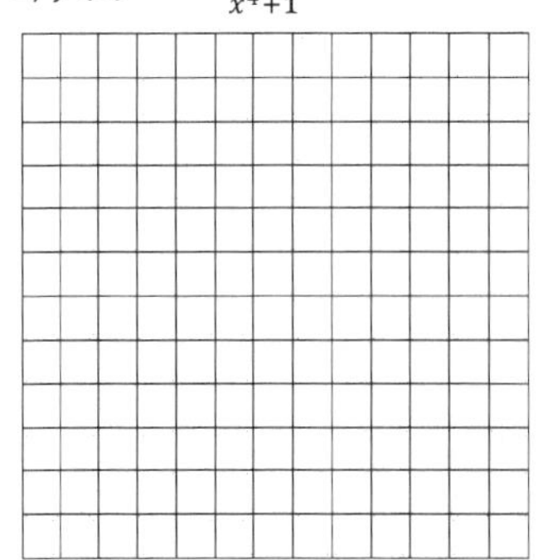

10) $f(x) = \frac{2x^2+3x-2}{x^3+1}$

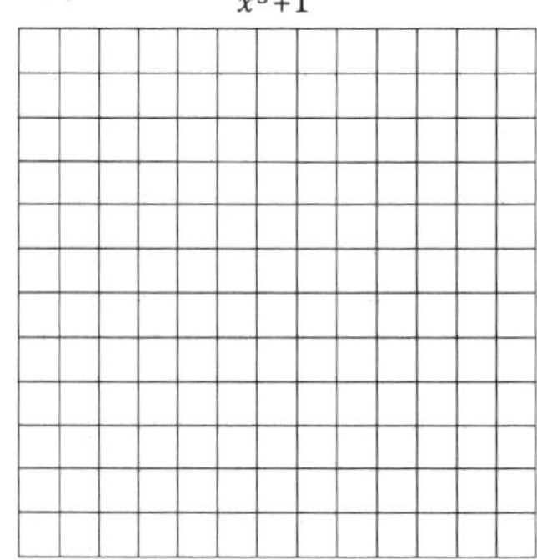

1	2	3	4	5	6	7	8	9	10
C	A	B	R	E	N	U	Z	O	M

2.D. Intermediate limits theorem

Theory and Examples

A continuous function is a function that has no sudden changes for all the values of the domain.

Intuitive EXAMPLE

Whenever we draw the graph of a continuous function, we don't have to take the pencil off the paper. As we draw the graph, the pencil remains on the paper for the entire graph.

The Intermediate limits theorem states in a nutshell, that for any closed interval of real numbers [a,b], a continuous function f(x) with the interval [a,b] as domain, and a point M between f(a) and f(b), $f(a) \neq f(b)$, there will be at least a point c between a and b for which f(c)=M.

EXAMPLE

As it can be seen in the graph below, the interval [a,b] is [-2,2]. The function is a continuous function of the relation:

$f(x) = 0.1x^3 + 1.5.$

There is a point M between f(-2)=0.7 and f(2)=2.3 for which the horizontal line will intersects the graph in a point of a value [c,f(c)] or [c,M].

In this case M is 1.4 and c=-1

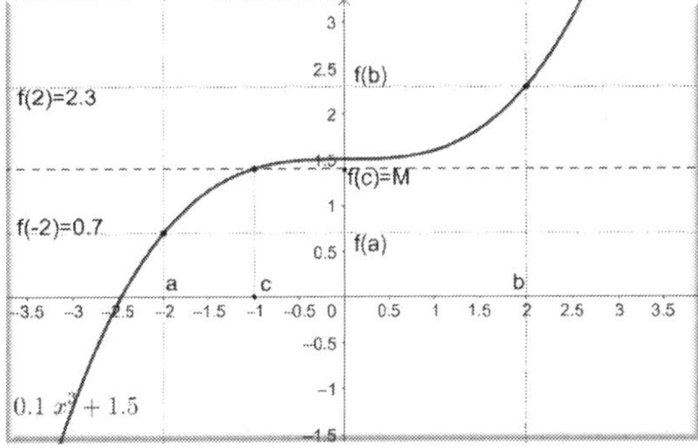

Chapter 2. D. Limits-Intermediate limits theorem

On which continent, other than North America, do elks live?

Determine which answer is correct. In the table at the bottom of the page cross off all the letters for the correct answers.

If f is a continuous function on the closed interval [-1,5] where $f(-1) = 2; f(5) = 7$

1) $f(c) = 6$ for at least one value of c that is between -1 and 5

2) $f(c) = 2$ for at least one value of c that is between -1 and 5

3) $f(c) = -2$ for at least one value of c that is between -1 and 5

4) $f(c) = 3$ for at least one value of c that is between 6 and 10

Using the table below of a continuous function determine if the following questions are correct.

x	-2	-1	0	3	5	7
$f(x)$	-1	0	3	4	5	9

5) The function $f(x) = 3$ for at least one x that belong to $-1 \leq x \leq 3$

6) The function $f(x) = 5$ for at least one x that belong to $-1 \leq x \leq 7$

7) The function $f(x) = 0$ for at least one x that belong to $0 \leq x \leq 3$

If $f(x) = \frac{2x+1}{-3x+7}$ determine if the questions below are correct.

8) There will be a value c such that $f(c) = 1$ for $3 \leq c \leq 9$

9) There will be a value c such that $f(c) = -2$ for $3 \leq c \leq 7$

10) There will be a value c such that $f(c) = 3$ for $-2 \leq c \leq 2$

1	2	3	4	5	6	7	8	9	10
C	R	A	S	T	O	I	A	V	P

2.E. Left and right limits

Theory and Examples

The left limit exists for values of x smaller than a certain value "a".

The notation for this limit is: $\lim_{x \to a^-} f(x)$

The right limit exists for values of x bigger than a certain value "a".

The notation for this limit is: $\lim_{x \to a^+} f(x)$

EXAMPLE

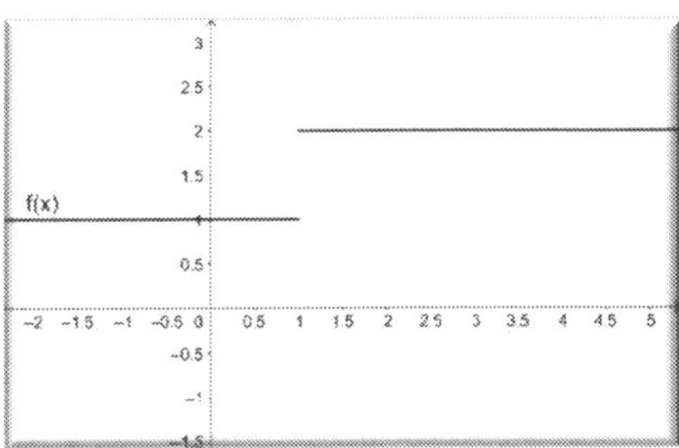

In this case the left limit of f(x) for values of x<1 is 1

$\lim_{x \to 1^-} f(x) = 1$

The right limit of f(x) for values of x>1 is 2

$\lim_{x \to 1^+} f(x) = 2$

As we can see, the limits are finite but are not the same. The limit for x=1 does not exist. In this case the function f(x) is not continuous in x=1.

EXAMPLE

The function in this case is:

$$f(x) = \begin{cases} 0.5x^3 + 1 \; if \; x < 0.5 \\ -x + 0.8 \; if \; x > 0.5 \end{cases}$$

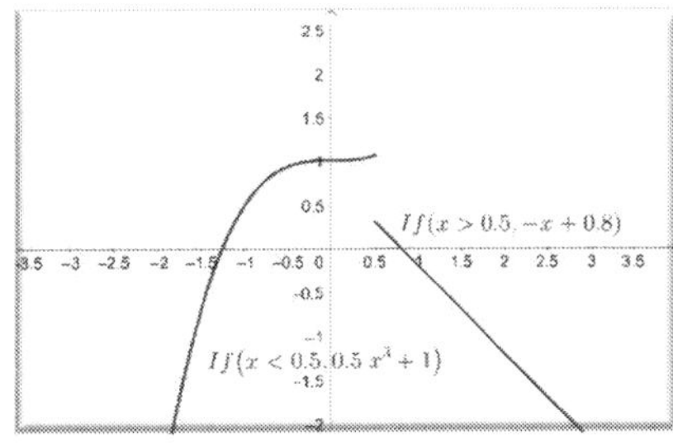

To find the left limit of the function around 0.5, we use the branch for which x values are smaller than 0.5.

$\lim_{x \to 0.5^-} f(x) = 0.5(0.5)^3 + 1 = 1.0625$

To find the right limit of the function around 0.5, we use the branch for which x values are bigger than 0.5.

$\lim_{x \to 0.5^+} f(x) = -0.5 + 0.8 = 0.3$

Chapter 2. E. Limits-Left and right limits

Which is the smallest of the bear species that has a short black fur?

Using the graph below, determine which answer is correct. In the table at the bottom of the page cross all the letters of the correct answer.

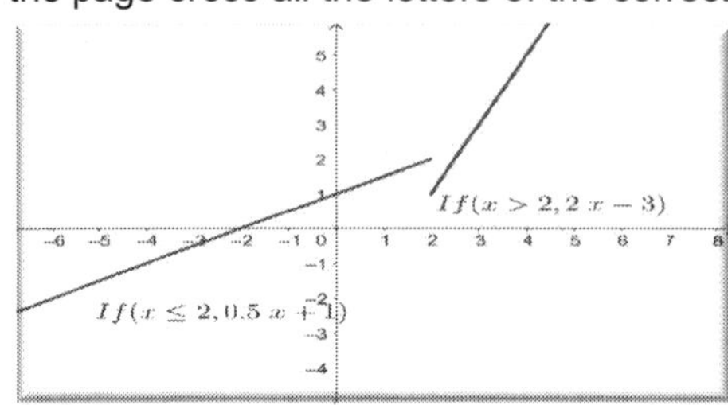

Where: $f(x) = \begin{cases} 0.5x + 1, x \leq 2 \\ 2x - 3, x > 2 \end{cases}$

1) $\lim_{x \to 0^-} f(x) = 1$

2) $\lim_{x \to 0^+} f(x) = 1$

3) $\lim_{x \to 2^-} f(x) = 2$

4) $\lim_{x \to 2^+} f(x) = 2$

Using the graph below, determine which answer is correct. In the table at the bottom of the page cross all the letters of the correct answer.

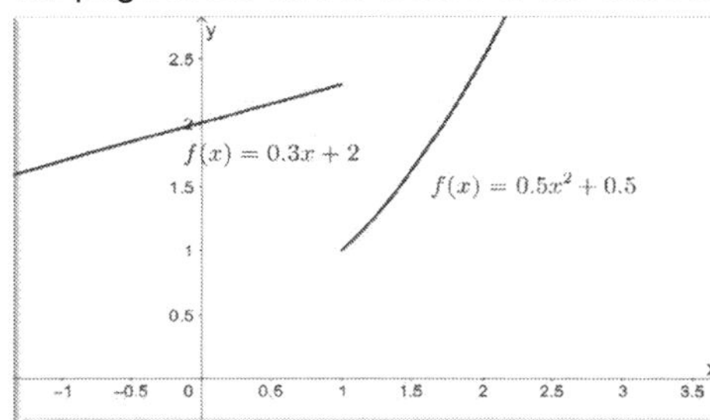

Where: $f(x) = \begin{cases} 0.3x + 2, x < 1 \\ 0.5x^2 + 0.5, x \geq 1 \end{cases}$

5) $\lim_{x \to 1^-} f(x) = 2.3$

6) $\lim_{x \to 1^+} f(x) = 1.5$

7) $\lim_{x \to 1} f(x) = does\ not\ exist$

8) $\lim_{x \to 3^-} f(x) = 3$

9) $\lim_{x \to 3^+} f(x) = 5$

10) $\lim_{x \to 7^-} f(x) = 25$

1	2	3	4	5	6	7	8	9	10
A	R	E	S	T	U	M	N	C	O

2.F. Limits to infinity

Theory and Examples

We have three cases when we have to find the limits of a rational expression algebraically for x approaches plus or minus infinity.

1. When the degree of the polynomial at the numerator is less than the degree of the polynomial at the denominator. In this case, the limit will always be zero for x going towards $\mp\infty$. The graph will have a horizontal asymptote x=0.

EXAMPLE

Find $\lim_{x\to\infty}\frac{x+1}{x^2-2}$

$$\lim_{x\to\infty}\frac{x+1}{x^2-2}=\lim_{x\to\infty}\frac{x+1}{x^2-2}=\lim_{x\to\infty}\left[\frac{x^2(\frac{1}{x}+\frac{1}{x^2})}{x^2(1-\frac{2}{x^2})}\right]=\frac{\lim_{x\to\infty}\left(\frac{1}{x}\right)+\lim_{x\to\infty}\left(\frac{1}{x^2}\right)}{\lim_{x\to\infty}1-\lim_{x\to\infty}\left(\frac{2}{x^2}\right)}=\frac{0+0}{1-0}=\frac{0}{1}=0$$

The graph of $f(x)=\frac{x+1}{x^2-2}$ is represented below.

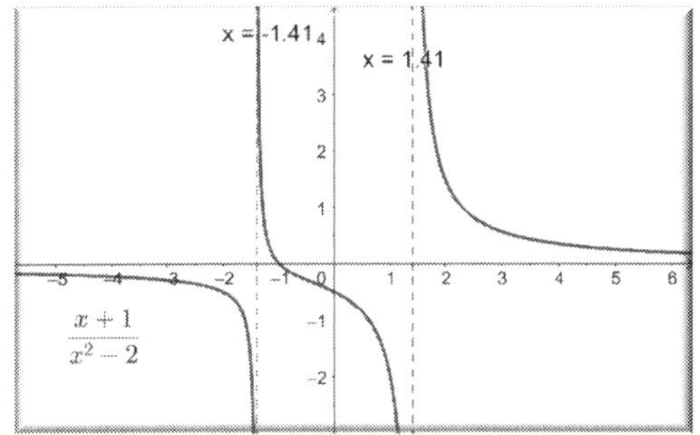

2. When the degree of the polynomial at the numerator is the same as the degree of the polynomial at the denominator. In this case, the limit will always equal the ratio between the coefficients of the terms with the highest exponent for x going towards $\mp\infty$. The graph will have a horizontal asymptote at x= the ratio between the coefficients of the terms with the highest exponent.

EXAMPLE

Find $\lim_{x\to\infty}\frac{3x^2+1}{x^2-2}$

$$\lim_{x\to\infty}\frac{3x^2+1}{x^2-2}=\lim_{x\to\infty}\frac{3x^2+1}{x^2-2}=\lim_{x\to\infty}\left[\frac{x^2(3+\frac{1}{x^2})}{x^2(1-\frac{2}{x^2})}\right]=\frac{\lim_{x\to\infty}(3)+\lim_{x\to\infty}\left(\frac{1}{x^2}\right)}{\lim_{x\to\infty}1-\lim_{x\to\infty}\left(\frac{2}{x^2}\right)}=\frac{3+0}{1-0}=\frac{3}{1}=3$$

The graph of $f(x) = \frac{3x^2+1}{x^2-2}$ is represented below.

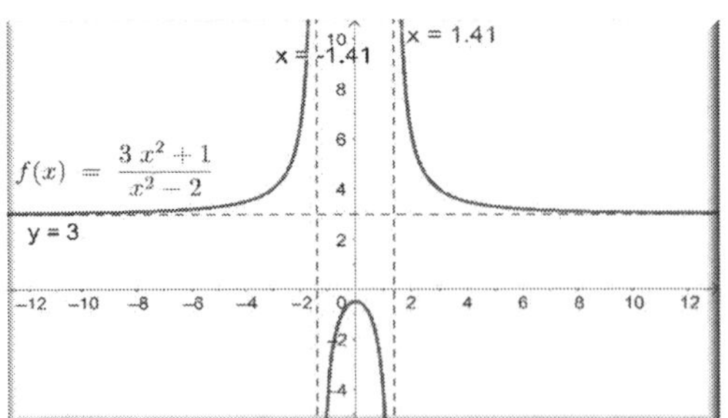

3. When the degree of the polynomial at the numerator is bigger than the degree of the polynomial at the denominator. In this case, the graph will tend to go either upward or downward, depending of the signs of the terms with the highest exponents from the numerator and denominator.

EXAMPLE

Find $\lim_{x\to\infty} \frac{3x^3+1}{x^2-2}$

$$\lim_{x\to\infty} \frac{3x^3+1}{x^2-2} = \lim_{x\to\infty} \frac{3x^3+1}{x^2-2} = \lim_{x\to\infty} [\frac{x^2(3x+\frac{1}{x^2})}{x^2(1-\frac{2}{x^2})}] = \frac{\lim_{x\to\infty}(3x)+\lim_{x\to\infty}(\frac{1}{x^2})}{\lim_{x\to\infty} 1-\lim_{x\to\infty}(\frac{2}{x^2})} = \frac{\infty+0}{1-0} = \infty$$

As we can see, both terms $3x^3$ and x^2 are positive. In this case, the result will be positive so, the graph will go toward plus infinity.

The graph of $f(x) = \frac{3x^3+1}{x^2-2}$ is represented below.

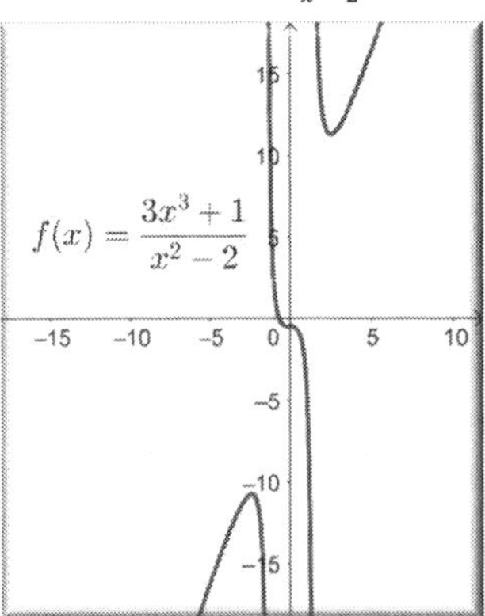

Chapter 2. F. Limits-Limits to infinity

Which is the name of the volcano that is the tallest in Japan and has 3776.24 m altitude?

Determine which answer is correct. In the table at the bottom of the page cross all the letters of the correct answer.

1) $\lim\limits_{x\to\infty} \frac{x^2+2x-45}{3x^3-5x^2+7x-3} = 0$

2) $\lim\limits_{x\to-\infty} \frac{x^5+5x-5}{2x^3-56x^2+8x-3} = \infty$

3) $\lim\limits_{x\to\infty} \frac{3x^2+2x-45}{3x^3-5x^2+7x-3} = 1$

4) $\lim\limits_{x\to-\infty} \frac{x-4}{2x^2-8x-2} = 0$

5) $\lim\limits_{x\to\infty} \frac{x^2+2x-45}{3x^2+37x-35} = \frac{1}{3}$

6) $\lim\limits_{x\to-\infty} \frac{3x^3+27x^2-4x+5}{2x^3-54x^2+27x-31} = 0$

7) $\lim\limits_{x\to\infty} \frac{9x^2-4}{5x^2+7x-3} = \frac{9}{5}$

8) $\lim\limits_{x\to\infty} \frac{3x+7}{\sqrt{x^2-3x+4}} = 0$

9) $\lim\limits_{x\to-\infty} \frac{5x-6}{\sqrt{x^2-5}} = 5$

10) $\lim\limits_{x\to-\infty} \frac{3x^2-2x+7}{3x^3-6x^2+8x-1} = 0$

11) $\lim\limits_{x\to\infty} \frac{2x+3}{\sqrt{x^3-27}} = 2$

12) $\lim\limits_{x\to\infty} \frac{x^2+2x-45}{4x^2-3x^2+7x-3} = \frac{1}{4}$

1	2	3	4	5	6
A	C	F	B	L	U
7	8	9	10	11	12
D	J	V	U	I	O

2.G. Continuity

Theory and Examples

Until now we had an intuitive approach to the continuity of functions. If we don't have to take the pencil off the page while we draw the function graph this function is continuous.

A rigorous definition of continuity says that:

a function f is continuous at a number c if,

$\lim_{x \to c} f(x) = f(c)$

The definition requires 3 things:

a) $f(c)$ is defined. It means that the value "c" belongs to the domain of f.

b) $\lim_{x \to c} f(x)$ exists

c) $\lim_{x \to c} f(x) = f(c)$

EXAMPLE

As it can be seen in the graph below, for a value c belonging to the domain of function f,

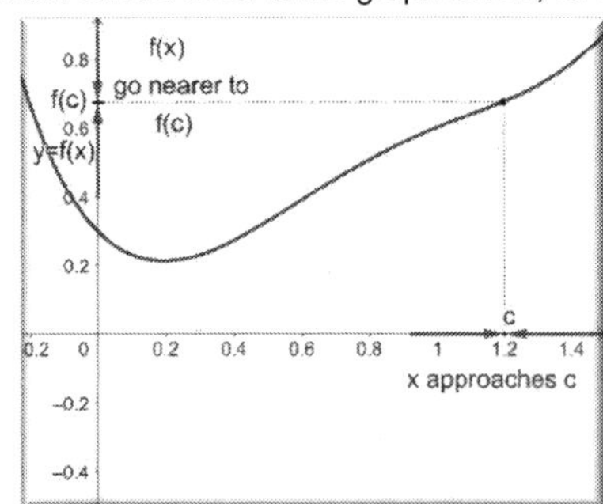

f(c) is defined. As x approaches "c" from left and right, the function f(x) approaches f(c).

$\lim_{x \to c} f(x)$ exists and equals f(c).

A function is <u>continuous coming from the left</u> of a value c if $\lim_{x \to c^-} f(x) = f(c)$

A function is <u>continuous coming from the right</u> of a value c if $\lim_{x \to c^+} f(x) = f(c)$

A function is **not** <u>continuous coming from the right or left</u> of a value c if,

$\lim_{x \to c^-} f(x) \neq \lim_{x \to c^+} f(x)$

EXAMPLE

For x approaching from the left, the left limit in the graph below is:

$\lim_{x \to 1^-} f(x) = f(1) = 2$

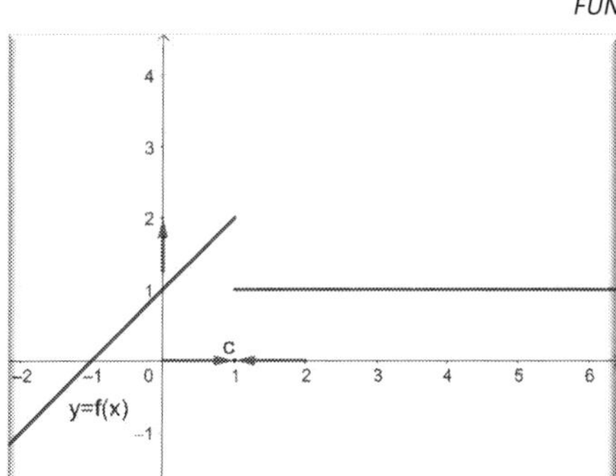

For x approaching from the right, the right limit in the graph below is $\lim_{x \to 1^+} f(x) = f(1) = 1$

The function exists for x=1,

The left and right limits exist but they are not the same.

In this case the function is not continuous for x=1.

Theorem

If functions f and g are continuous at c and b is a constant, the following functions are continuous as well.

f+g ; f-g ; bf ; fg ; f/g if $g(c) \neq 0$

Theorem

1. Any polynomial function is continuous everywhere (on real numbers).
2. Any rational function is continuous wherever it is defined; it is continuous on its domain.
3. These types of functions are continuous on their domain as well.

   Root functions and trigonometric functions are continuous functions.

Theorem

If f is continuous at c and $\lim_{x \to c} g(x) = c$ then $\lim_{x \to c} f[g(x)] = f(c)$

or;

$\lim_{x \to c} f[g(x)] = f[\lim_{x \to c} g(x)]$

Chapter 2. G. Limits-Continuity

In 83 BC the Dictator of Rome asked young Julius Caesar to divorce his beloved wife Cornelia. Young Caesar disobeyed this formidable and dangerous Dictator of Rome and left Rome in exile until the Dictator's death in 78 BC. Who was this Dictator of Rome?

Determine which answer is correct. In the table at the bottom of the page cross all the letters of the correct answer.

The following function is represented below: $f(x) = \begin{cases} 2x-1, x \geq 2 \\ 3x-3, 1.5 < x < 2 \\ 0.5x, x \leq 1.5 \end{cases}$

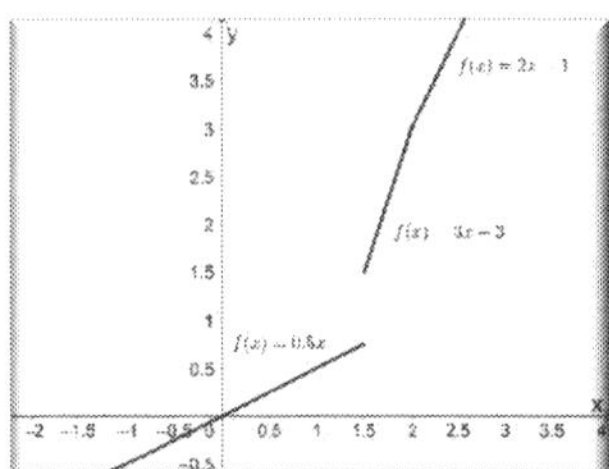

1) The function f(x) is continuous in x=4

2) The function f(x) is not continuous in x=2

3) The function f(x) is continuous in x=1

4) The function f(x) is continuous in x=-7

5) The function f(x) is continuous in x=1.5

The function g(x) is defined as follows: $g(x) = \begin{cases} \frac{9x^2-1}{3x-1} \; for \; x \neq 1/3 \\ 0 \; for \; x = 1/3 \end{cases}$

6) The function g(x) exists for x=1/3

7) The limit $\lim_{x \to 1/3} g(x)$ does not exist.

8) The function g(x) is continuous for x=0.33

9) The function $f(x) = \begin{cases} 3x \; if -1 \leq x \leq 2 \\ x+2 \; if \; x > 2 \end{cases}$ is continuous over interval [-1,2]

10) The function $f(x) = \begin{cases} 5x \; if -1 \leq x \leq 3 \\ 2x+2 \; if \; x > 3 \end{cases}$ is continuous over [-5,5]

1	2	3	4	5	6	7	8	9	10
E	S	R	I	U	H	L	L	O	A

QUICK ANSWERS

Chapter 2

2.A.a	BUMP
2.A.b	BRUTUS
2.B	MARE
2.C	BENZ
2.D	ASIA
2.E	SUN
2.F	FUJI
2.G	SULLA

FULL SOLUTIONS

CHAPTER 2

Chapter 2. A. a. Limits – Table of values, and graphically

1. Correct

We can show limits through graphs or tables of values.

2. Incorrect

If x=2 doesn't belong to the domain, the function could have limit for f(2).

3. Correct

The table below represent the value of $f(x) = 0.3x^2 - 2$ around -0.8 for x around 2

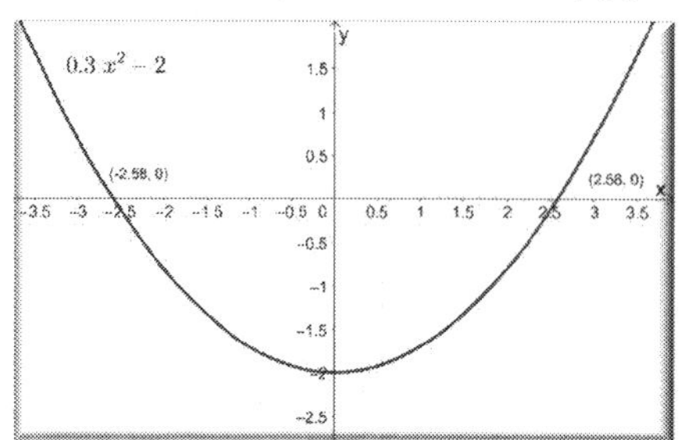

X	$f(x) = 0.3x^2 - 2$
1.8	-1.028
1.9	-0.917
1.999	-0.8012
2.001	-0.7988
2.01	-0.7879
2.1	-0.677

4. Incorrect

As x is approaching 5, f(x) is approaching 28

X	$f(x) = x^2 + 3$
4.8	26.04
4.99	27.9
4.999	27.99
5.001	28.01
5.01	28.1
5.1	29.1

5. Incorrect

The value of $f(x) = x^2 + 3$ for x around 5 is not around 7.6. When x approaches 5 the value of the function goes towards 28.

6. Incorrect

The table below represents the value of $f(x) = x^2 + 2x$ around 8 for x around 2.

X	$f(x) = x^2 + 2x$
1.89	7.35
1.999	7.98
2.001	8.006
2.01	8.06
2.1	8.61

Chapter 2. A. b. Limits – Algebraically

1. Incorrect

$$\lim_{x\to 0}\left(\frac{6x^2-3}{x+3}\right) = \frac{\lim_{x\to 0}(6x^2-3)}{\lim_{x\to 0}(x+3)} = \frac{-3}{3} = -1$$

2. Correct

$$\lim_{x\to 3}\left(\frac{9x^2-1}{3x-1}\right) = \lim_{x\to 3}\left[\frac{(3x-1)(3x+1)}{(3x-1)}\right] = \lim_{x\to 3}(3x+1) = 10$$

3. Incorrect

$$\lim_{x\to 2}\frac{x^2+2x-8}{x-2} = \lim_{x\to 2}\left[\frac{(x-2)(x+4)}{x-2}\right] = \lim_{x\to 2}(x+4) = 6$$

4. Incorrect

$$\lim_{x\to 3}\left(\frac{1}{x-3} - \frac{6}{x^2-9}\right) = \lim_{x\to 3}\left[\frac{1}{x-3}\left(1 - \frac{6}{x+3}\right)\right] = \lim_{x\to 3}\left[\frac{1}{x-3}\left(\frac{x+3-6}{x+3}\right)\right] = \lim_{x\to 3}\left[\frac{1}{x-3}\left(\frac{x-3}{x+3}\right)\right] = \lim_{x\to 3}\frac{1}{x+3} = \frac{1}{6}$$

5. Correct

$$\lim_{x\to\infty}\frac{4x+3}{x^2+4x-2} = \lim_{x\to\infty}\left[\frac{x^2\left(\frac{4}{x}+\frac{3}{x^2}\right)}{x^2\left(1+\frac{4}{x}-\frac{2}{x^2}\right)}\right] = \frac{\lim_{x\to\infty}\left(\frac{4}{x}\right)+\lim_{x\to\infty}\left(\frac{3}{x^2}\right)}{\lim_{x\to\infty}1+\lim_{x\to\infty}\left(\frac{4}{x}\right)-\lim_{x\to\infty}\left(\frac{2}{x^2}\right)} = \frac{0+0}{1+0+0} = \frac{0}{1} = 0$$

6. Incorrect

$$\lim_{x\to\infty}\left(\sqrt{2x+1} - \sqrt{2x}\right) = \lim_{x\to\infty}\left(\sqrt{2x+1} - \sqrt{2x}\right)\left(\frac{\sqrt{2x+1}+\sqrt{2x}}{\sqrt{2x+1}+\sqrt{2x}}\right) = \lim_{x\to\infty}\left(\frac{2x+1-2x}{\sqrt{2x+1}+\sqrt{x}}\right) = \lim_{x\to\infty}\frac{1}{\sqrt{2x+1}+\sqrt{x}} = 0$$

7. Correct

$\lim_{x\to 3}(\frac{16x^2-1}{4x-1}) = \lim_{x\to 3}\left[\frac{(4x-1)(4x+1)}{4x-1}\right] = \lim_{x\to 3}(4x+1) = 13$

8. Correct

$\lim_{x\to 1}\left(\frac{x^4-1}{x-1}\right) = \lim_{x\to 1}\left[\frac{(x^2-1)(x^2+1)}{x-1}\right] = \lim_{x\to 1}\left[\frac{(x-1)(x+1)(x^2+1)}{x-1}\right] = \lim_{x\to 1}(x+1)(x^2+1) = \lim_{x\to 1}(x+1) * \lim_{x\to 1}(x^2+1) = 2 * 2 = 4$

9. Incorrect

$\lim_{x\to 1}\left(\frac{x^4-1}{x^2-1}\right) = \lim_{x\to 1}\left[\frac{(x^2-1)(x^2+1)}{(x-1)(x+1)}\right] = \lim_{x\to 1}\left[\frac{(x-1)(x+1)(x^2+1)}{(x-1)(x+1)}\right] = \lim_{x\to 1}(x^2+1) = 2$

10. Incorrect

$\lim_{h\to 0}\frac{f(x+h)-f(x)}{h}$ $when\ f(x) = 3x^2 + 4x$

$f(x+h) = 3(x+h)^2 + 4(x+h) = 3(x^2+2xh+h^2) + 4x + 4h = 3x^2 + 6xh + 3h^2 + 4x + 4h$

$f(x) = 3x^2 + 4x$

So: $f(x+h) - f(x) = 3x^2 + 6xh + 3h^2 + 4x + 4h - (3x^2 + 4x) = 3x^2 + 6xh + 3h^2 + 4x + 4h - 3x^2 - 4x = 6xh + 4h + 3h^2$

Then, $\frac{f(x+h)-f(x)}{h} = \frac{6xh+4h+3h^2}{h} = \frac{h(6x+4+3h)}{h} = 6x + 4 + 3h$

So $\lim_{h\to 0}\frac{f(x+h)-f(x)}{h} = \lim_{h\to 0}(6x + 4 + 3h) = 6x + 4$

Chapter 2. B. Limits-One side versus two sides

In the graph below $g(x) = -(0.5x-3)^2 + 3\ for\ x < 5$

$g(x) = (0.4x-3)^2 + 1\ for\ x \geq 5$

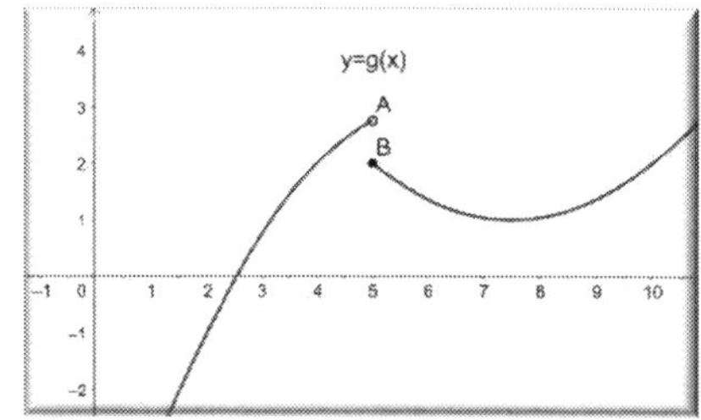

1. Correct

$\lim_{x\to 5^-} g(x) = 2.75$

2. Correct

$\lim_{x\to 5^+} g(x) = 2$

3. Incorrect

As $\lim_{x\to 5^-} g(x) = 2.75$ and $\lim_{x\to 5^+} g(x) = 2$ $\lim_{x\to 5} g(x) \neq 5$

The limit does not exist because, there are two different values of g(x) when x gets near to 5.

In the graph below

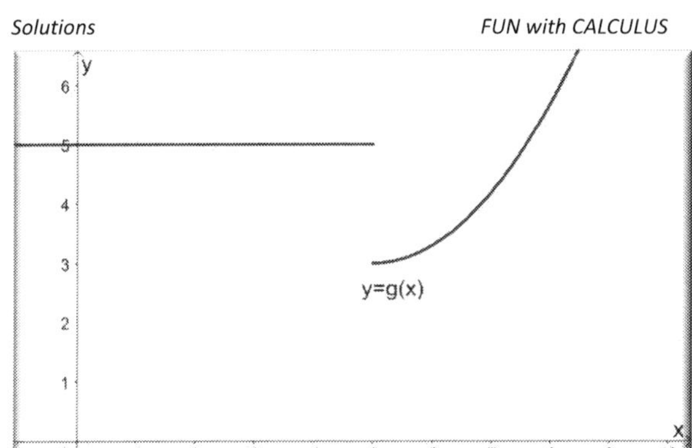

4. Correct

$\lim_{x\to5^-} g(x) = 5$

5. Correct

$\lim_{x\to5^+} g(x) = 3$

6, Incorrect

As $\lim_{x\to5^-} g(x) = 6$ and $\lim_{x\to5^+} g(x) = 2$

$\lim_{x\to5} g(x) \neq 5$

The limit does not exist because, there are two different values of g(x) when x gets near to 5.

7. Correct

$$\lim_{x\to5^+} \frac{5x}{x-5} = \frac{25}{positive\ very\ small\ number} = +\infty$$

8. Incorrect

$$\lim_{x\to5^-} \frac{5x}{x-4} = \lim_{x\to5^-} \frac{5(5)}{1} = 25$$

9. Correct

$$\lim_{x\to3^-} \frac{5x}{x-5} = \frac{\lim_{x\to3^-} 5x}{\lim_{x\to3^-}(x-5)} = \frac{15}{-2} = -7.5$$

10. Incorrect

$$\lim_{x\to3^-} \frac{x}{x-5} = -1.5$$

$$\lim_{x\to3^-} \frac{x}{x-5} = \frac{\lim_{x\to3^-} x}{\lim_{x\to3^-}(x-5)} = \frac{3}{-2} = -1.5$$

Chapter 2. C. Limits-End behavior

Determine what is the end behavior of these functions using limits.

1. Correct

$f(x) = 2x^4 - 3x^2 + 15x - 24$: when x goes to $+\infty$ it goes up in first quadrant.
Point a) in the Theory section. The polynomial function has an even degree and the coefficient of the highest degree term is positive.

2. Correct

$f(x) = -4x^4 + 2x^2 - 5x$: when x goes to -∞ it goes down in the third quadrant.
Point b) in the Theory section. The polynomial function has an even degree and the coefficient of the highest degree term is negative.

3. Incorrect

$f(x) = -4x^3 + 6x^2 - x$: when x goes to -∞ it goes up in the second quadrant.
Point c) in the Theory section. The polynomial function has an odd degree and the coefficient of the highest degree term is negative.

4. Correct

$f(x) = x^3 - 36x^2 - x + 73$: when x goes to -∞ it goes down in the third quadrant
Point c) in the Theory section. The polynomial function has an odd degree and the coefficient of the highest degree term is positive.

5. Incorrect

$f(x) = \frac{2x^3+3x^2-4x+5}{6x^3-5x+4}$: when x goes to +∞ $f(x)$ goes towards $\frac{1}{3}$

$$\lim_{x\to\infty} \frac{2x^3+3x^2-4x+5}{6x^3-5x+4} = \lim_{x\to\infty} \frac{x^3(2+\frac{3}{x}-\frac{4}{x^2}+\frac{5}{x^3})}{x^3(6-\frac{5}{x^2}+\frac{4}{x^3})} = \lim_{x\to\infty} \frac{(2+\frac{3}{x}-\frac{4}{x^2}+\frac{5}{x^3})}{(6-\frac{5}{x^2}+\frac{4}{x^3})} = \frac{\lim_{x\to\infty}(2+\frac{3}{x}-\frac{4}{x^2}+\frac{5}{x^3})}{\lim_{x\to\infty}(6-\frac{5}{x^2}+\frac{4}{x^3})} = \frac{2+0-0+0}{6-0+0} = \frac{2}{6} = \frac{1}{3}$$

6. Incorrect

$f(x) = \frac{x^3+2x^2-3x+4}{6x^4-5x^3+4x^2-3x}$: when x goes to -∞ $f(x)$ goes towards 0

$$\lim_{x\to-\infty} \frac{x^3+2x^2-3x+4}{6x^4-5x^3+4x^2-3x} = \lim_{x\to-\infty} \frac{x^3(1+\frac{2}{x}-\frac{3}{x^2}+\frac{4}{x^3})}{x^4(6-\frac{5}{x}+\frac{4}{x^2}-\frac{3}{x^3})} = \lim_{x\to-\infty} \frac{(1+\frac{2}{x}-\frac{3}{x^2}+\frac{4}{x^3})}{x(6-\frac{5}{x}+\frac{4}{x^2}-\frac{3}{x^3})} = \frac{\lim_{x\to-\infty}(1+\frac{2}{x}-\frac{3}{x^2}+\frac{4}{x^3})}{\lim_{x\to-\infty} x(6-\frac{5}{x}+\frac{4}{x^2}-\frac{3}{x^3})} = \frac{1+0-0+0}{\infty*(6-0+0-0)} = 0$$

7. Correct

$f(x) = \frac{x^5+2x^2-3x+4}{6x^4-5x^3+4x^2-3x}$: when x goes to +∞ $f(x)$ goes up in the first quadrant.
When the degree of the polynomial at the numerator is bigger than the degree of the polynomial at the denominator, for x going towards plus infinite, the graph goes upward in the first quadrant.
The graph is represented below.

8. Incorrect

$f(x) = \frac{2x^3+3x^2-4x+5}{6x^2-5x+4}$: when x goes to -∞ it goes down in the third quadrant.

When the degree of the polynomial at the numerator is bigger than the degree of the polynomial at the denominator, for x going towards minus infinite, the graph goes downward in the third quadrant. The graph is represented below.

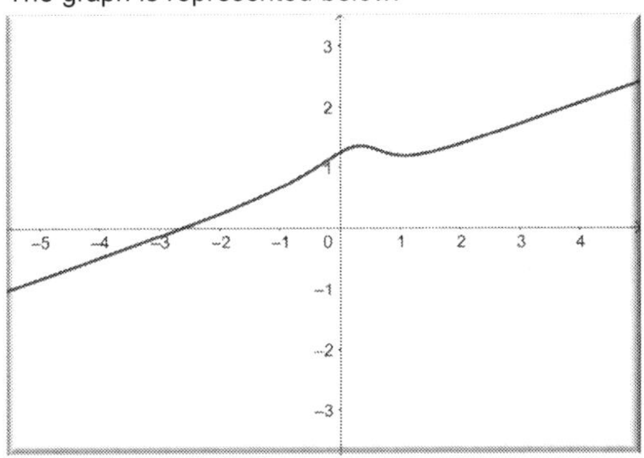

Graph the following functions and determine the end behavior for x going towards +∞

9.

F(x) is going towards 0.

$f(x) = \frac{x^3-2x^2+3x}{x^4+1}$

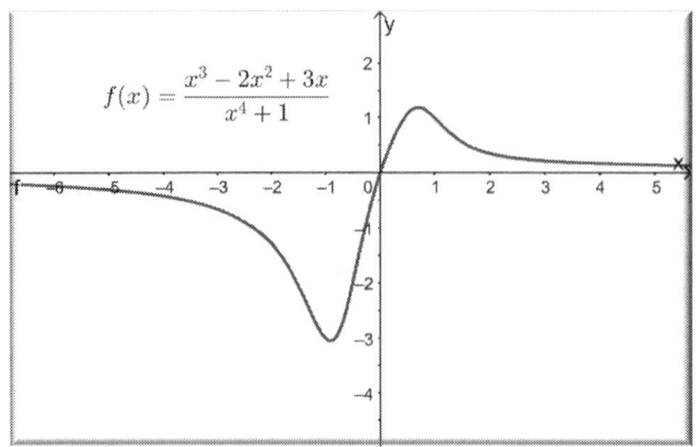

10.

F(x) is going towards 0.

$$f(x) = \frac{2x^2 + 3x - 2}{x^3 + 1}$$

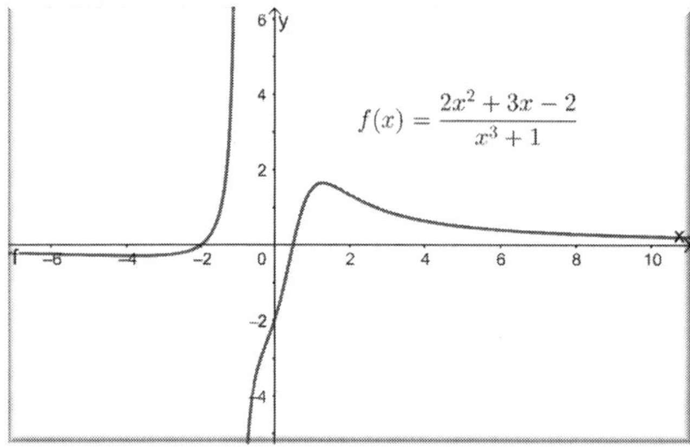

Chapter 2. D. Limits-Intermediate limits theorem

1. Correct

$f(c) = 6$ for at least one value of c that is between -1 and 5

2. Correct

$f(c) = 2$ for at least one value of c that is between -1 and 5

3. Incorrect

$f(c) = -2$ for at least one value of c that is between -1 and 5
-2 does not belong between 2 and 7

4). Incorrect

$f(c) = 3$ for at least one value of c that is between 6 and 10
6 and 10 are outside the closed interval [-1,5]
Using the table below of a continuous function determine if the following questions are correct.

x	-2	-1	0	3	5	7
$f(x)$	-1	0	3	4	5	9

5. Correct

The function $f(x) = 3$ for at least one x that belong to $-1 \le x \le 3$
In this case x=0

6. Correct

The function $f(x) = 5$ for at least one x that belong to $-1 \le x \le 7$
In this case x=5

7. Incorrect

The function $f(x) = 0$ for at least one x that belong to $0 \le x \le 3$
$f(x) = 0$ for x=-1. X=-1 is outside the interval [0,3]
If $f(x) = \frac{2x+1}{-3x+7}$ determine if the questions below are correct.

8. Incorrect

There will not be a value c such that $f(c) = 1$ for $3 \le c \le 9$
So $f(3) = \frac{2(3)+1}{-3(3)+7} = \frac{7}{-2} = -3.5; f(9) = \frac{2(9)+1}{-3(9)+7} = \frac{19}{-20} = -0.95$
F(x) is always between -3.5 and -0.95
There is no value of c between 3 and 9 for which $f(c) = 1$

9. Correct

There will be a value c such that $f(c) = -2$ for $3 \le c \le 7$
So $f(3) = \frac{2(3)+1}{-3(3)+7} = \frac{7}{-2} = -3.5\,;\ f(7) = \frac{2(7)+1}{-3(7)+7} = \frac{15}{-14} = -1.07$
F(x) is always between -3.5 and -1.07

Yes, there will be a value c such that $f(c) = -2$ for $3 \leq c \leq 7$

10. Correct

There will be a value c such that $f(c) = 2$ for $-2 \leq c \leq 2$

So $f(-2) = \frac{2(-2)+1}{-3(-2)+7} = \frac{-3}{13} = -0.23; f(2) = \frac{2(2)+1}{-3(2)+7} = \frac{5}{1} = 5$

Yes, there will be a value c such that $f(c) = 2$ for $-2 \leq c \leq 2$

Chapter 2. E. Limits-Left and right limits

Using the graph below, determine which answer is correct. In the table at the bottom of the page cross all the letters of the correct answer.

Where: $f(x) = \begin{cases} 0.5x + 1, x \leq 2 \\ 2x - 3, x > 2 \end{cases}$

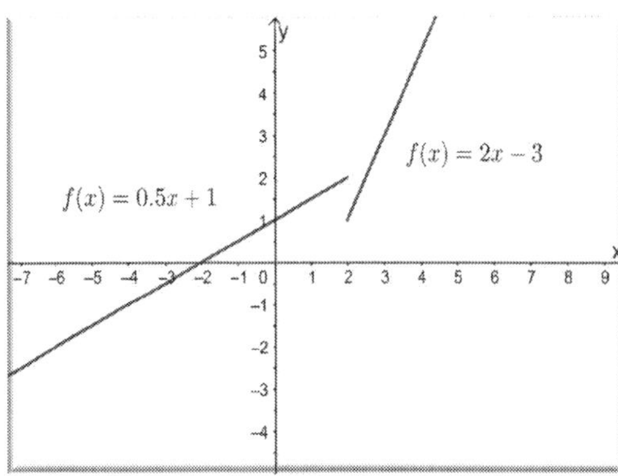

1. Correct

$\lim_{x \to 0^-} f(x) = 0.5(0) + 1 = 1$

2. Correct

$\lim_{x \to 0^+} f(x) = 1$

3. Correct

$\lim_{x \to 2^-} f(x) = 0.5(2) + 1 = 1 + 1 = 2$

4. Incorrect

$\lim_{x \to 2^+} f(x) = 2(2) - 3 = 4 - 3 = 1$

Using the graph below, determine which answer is correct.

Where: $f(x) = \begin{cases} 0.3x + 2, x < 1 \\ 0.5x^2 + 0.5, x \geq 1 \end{cases}$

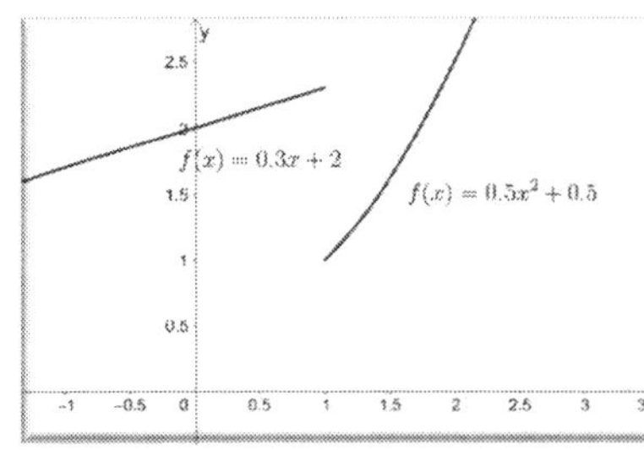

5. Correct

$\lim_{x \to 1^-} f(x) = 0.3(1) + 2 = 0.3 + 2 = 2.3$

6. Incorrect

$\lim_{x \to 1^+} f(x) = 0.5(1)^2 + 0.5 = 0.5 + 0.5 = 1$

7. Correct

As $\lim_{x \to 1^-} f(x) = 2.3$ and $\lim_{x \to 1^+} f(x) = 1$ The limit does not exist because, there are two different values of f(x) when x gets near to 1.

8. Incorrect

$$\lim_{x\to 3^-} f(x) = 0.5(3)^2 + 0.5 = 0.5(9) + 0.5 = 4.5 + 0.5 = 5$$

9. Correct

$$\lim_{x\to 3^+} f(x) = 5$$

10. Correct

$$\lim_{x\to 7^-} f(x) = 0.5(7)^2 + 0.5 = 0.5(49) + 0.5 = 24.5 + 0.5 = 25$$

Chapter 2. F. Limits-Limits to infinity

1. Correct

$$\lim_{x\to\infty} \frac{x^2+2x-45}{3x^3-5x^2+7x-3} = \lim_{x\to\infty} \frac{x^2(1+\frac{2}{x}-\frac{45}{x^2})}{x^3(3-\frac{5}{x}+\frac{7}{x^2}-\frac{3}{x^3})} = \lim_{x\to\infty} \frac{(1+\frac{2}{x}-\frac{45}{x^2})}{x(3-\frac{5}{x}+\frac{7}{x^2}-\frac{3}{x^3})} = \frac{\lim_{x\to\infty}(1)+\lim_{x\to\infty}(\frac{2}{x})-\lim_{x\to\infty}(\frac{45}{x^2})}{\lim_{x\to\infty}(x)[\lim_{x\to\infty}(3)-\lim_{x\to\infty}\left(\frac{5}{x}\right)+\lim_{x\to\infty}\left(\frac{7}{x^2}\right)-\lim_{x\to\infty}\left(\frac{3}{x^3}\right)]} =$$

$$\frac{1+0+0}{\lim_{x\to\infty}(x)(3-0+0-0)} = 0$$

2. Correct

$$\lim_{x\to-\infty} \frac{x^5+5x-5}{2x^3-56x^2+8x-3} = \lim_{x\to-\infty} \frac{x^5(1+\frac{5}{x^4}-\frac{5}{x^5})}{x^3(2-\frac{56}{x}+\frac{8}{x^2}-\frac{3}{x^3})} = \lim_{x\to-\infty} \frac{x^2(1+\frac{5}{x^4}-\frac{5}{x^5})}{(2-\frac{56}{x}+\frac{8}{x^2}-\frac{3}{x^3})} = \frac{\lim_{x\to-\infty}(x^2)[\lim_{x\to-\infty}(1)+\lim_{x\to-\infty}\left(\frac{5}{x^4}\right)-\lim_{x\to-\infty}\left(\frac{5}{x^5}\right)]}{\lim_{x\to-\infty}(2)-\lim_{x\to-\infty}(\frac{56}{x})+\lim_{x\to-\infty}(\frac{8}{x^2})-\lim_{x\to-\infty}(\frac{3}{x^3})} =$$

$$\frac{\lim_{x\to-\infty}(x^2)(1+0-0)}{2-0+0-0} = \infty$$

3. Incorrect

$$\lim_{x\to\infty} \frac{3x^2+2x-45}{3x^3-5x^2+7x-3} = \lim_{x\to\infty} \frac{x^2(3+\frac{2}{x}-\frac{45}{x^2})}{x^3(3-\frac{5}{x}+\frac{7}{x^2}-\frac{3}{x^3})} = \lim_{x\to\infty} \frac{(3+\frac{2}{x}-\frac{45}{x^2})}{x(3-\frac{5}{x}+\frac{7}{x^2}-\frac{3}{x^3})} \lim_{x\to\infty} \frac{\lim_{x\to\infty}(3)+\lim_{x\to\infty}(\frac{2}{x})-\lim_{x\to\infty}(\frac{45}{x^2})}{\lim_{x\to\infty}(x)[\lim_{x\to\infty}(3)-\lim_{x\to\infty}\left(\frac{5}{x}\right)+\lim_{x\to\infty}\left(\frac{7}{x^2}\right)-\lim_{x\to\infty}\left(\frac{3}{x^3}\right)]} =$$

$$\frac{3+0-0}{\lim_{x\to\infty}(x)(3-0+0-0)} = \frac{3}{\lim_{x\to\infty}(x)(3)} = 0$$

4. Correct

$$\lim_{x\to-\infty} \frac{x-4}{2x^2-8x-2} = \lim_{x\to-\infty} \frac{x(1-\frac{4}{x})}{x^2(2-\frac{8}{x}-\frac{2}{x^2})} = \lim_{x\to-\infty} \frac{(1-\frac{4}{x})}{x(2-\frac{8}{x}-\frac{2}{x^2})} = \frac{\lim_{x\to-\infty}(1)-\lim_{x\to-\infty}(\frac{4}{x})}{\lim_{x\to-\infty}(x)[\lim_{x\to-\infty}(2)-\lim_{x\to-\infty}\left(\frac{8}{x}\right)-\lim_{x\to-\infty}\left(\frac{2}{x^2}\right)]} = \frac{1-0}{\lim_{x\to-\infty}(x)(2-0-0)} =$$

$$\frac{1}{\lim_{x\to-\infty}(x)(2)} = 0$$

5. Correct

$$\lim_{x\to\infty} \frac{x^2+2x-45}{3x^2+37x-35} = \lim_{x\to\infty} \frac{x^2(1+\frac{2}{x}-\frac{45}{x^2})}{x^2(3+\frac{37}{x}-\frac{35}{x^2})} = \lim_{x\to\infty} \frac{(1+\frac{2}{x}-\frac{45}{x^2})}{(3+\frac{37}{x}-\frac{35}{x^2})} = \frac{\lim_{x\to\infty}(1)+\lim_{x\to\infty}(\frac{2}{x})-\lim_{x\to\infty}(\frac{45}{x^2})}{\lim_{x\to\infty}(3)+\lim_{x\to\infty}\frac{37}{x})-\lim_{x\to\infty}(\frac{35}{x^2})} = \frac{1+0-0}{3+0-0} = \frac{1}{3}$$

6. Incorrect

$$\lim_{x\to-\infty}\frac{3x^3+27x^2-4x+5}{2x^3-54x^2+27x-31}=\lim_{x\to-\infty}\frac{x^3(3+\frac{27}{x}-\frac{4}{x^2}+\frac{5}{x^3})}{x^3(2-\frac{54}{x}+\frac{27}{x^2}-\frac{31}{x^3})}=\lim_{x\to-\infty}\frac{(3+\frac{27}{x}-\frac{4}{x^2}+\frac{5}{x^3})}{(2-\frac{54}{x}+\frac{27}{x^2}-\frac{31}{x^3})}=\frac{\lim_{x\to-\infty}(3)+\lim_{x\to-\infty}(\frac{27}{x})-\lim_{x\to-\infty}(\frac{4}{x^2})+\lim_{x\to-\infty}(\frac{5}{x^3})}{\lim_{x\to-\infty}(2)-\lim_{x\to-\infty}(\frac{54}{x})+\lim_{x\to-\infty}(\frac{27}{x^2})-\lim_{x\to-\infty}(\frac{31}{x^3})}=$$

$$\frac{3+0-0+0}{2-0+0-0}=\frac{3}{2}$$

7. Correct

$$\lim_{x\to\infty}\frac{9x^2-4}{5x^2+7x-3}=\lim_{x\to\infty}\frac{x^2(9-\frac{4}{x^2})}{x^2(5+\frac{7}{x}-\frac{4}{x^2})}=\lim_{x\to\infty}\frac{(9-\frac{4}{x^2})}{(5+\frac{7}{x}-\frac{4}{x^2})}=\frac{\lim_{x\to\infty}(9)-\lim_{x\to\infty}(\frac{4}{x^2})}{\lim_{x\to\infty}(5)+\lim_{x\to\infty}(\frac{7}{x})-\lim_{x\to\infty}(\frac{4}{x^2})}=\frac{9-0}{5+0-0}=\frac{9}{5}$$

8. Incorrect

$$\lim_{x\to\infty}\frac{3x+7}{\sqrt{x^2-3x+4}}=\lim_{x\to\infty}\frac{x(3+\frac{7}{x})}{\sqrt{x^2(1-\frac{3}{x}+\frac{4}{x^2})}}=\lim_{x\to\infty}\frac{x(3+\frac{7}{x})}{x\sqrt{(1-\frac{3}{x}+\frac{4}{x^2})}}=\lim_{x\to\infty}\frac{(3+\frac{7}{x})}{\sqrt{(1-\frac{3}{x}+\frac{4}{x^2})}}=\frac{\lim_{x\to\infty}(3)+\lim_{x\to\infty}(\frac{7}{x})}{\sqrt{\lim_{x\to\infty}(1)-\lim_{x\to\infty}\frac{3}{x}+\lim_{x\to\infty}(\frac{4}{x^2})}}=\frac{3+0}{\sqrt{1-0+0}}=\frac{3}{1}=3$$

9. Correct

$$\lim_{x\to-\infty}\frac{5x-6}{\sqrt{x^2-5}}=\lim_{x\to-\infty}\frac{x(5-\frac{6}{x})}{\sqrt{x^2(1-\frac{5}{x^2})}}=\lim_{x\to-\infty}\frac{x(5-\frac{6}{x})}{x\sqrt{(1-\frac{5}{x^2})}}=\lim_{x\to-\infty}\frac{(5-\frac{6}{x})}{\sqrt{(1-\frac{5}{x^2})}}=\frac{\lim_{x\to-\infty}(5)-\lim_{x\to-\infty}(\frac{6}{x})}{\sqrt{\lim_{x\to-\infty}(1)-\lim_{x\to-\infty}(\frac{5}{x^2})}}=\frac{5-0}{\sqrt{1-0}}=5$$

10. Correct

$$\lim_{x\to-\infty}\frac{3x^2-2x+7}{3x^3-6x^2+8x-1}=\lim_{x\to-\infty}\frac{x^2(3-\frac{2}{x}+\frac{7}{x^2})}{x^3(3-\frac{6}{x}+\frac{8}{x^2}-\frac{1}{x^3})}=\lim_{x\to-\infty}\frac{(3-\frac{2}{x}+\frac{7}{x^2})}{x(3-\frac{6}{x}+\frac{8}{x^2}-\frac{1}{x^3})}=\frac{\lim_{x\to-\infty}(3)-\lim_{x\to-\infty}\left(\frac{2}{x}\right)+\lim_{x\to-\infty}\frac{7}{x^2}}{\lim_{x\to-\infty}(x)[\lim_{x\to-\infty}(3)-\lim_{x\to-\infty}(\frac{6}{x})+\lim_{x\to-\infty}(\frac{8}{x^2})-\lim_{x\to-\infty}(\frac{1}{x^3})]}=$$

$$\frac{3-0+0}{\lim_{x\to-\infty}(x)(3-0+0-0)}=\frac{3}{\lim_{x\to-\infty}(x)(3)}=0$$

11. Incorrect

$$\lim_{x\to\infty}\frac{2x+3}{\sqrt{x^3-27}}=\lim_{x\to\infty}\frac{x(2+\frac{3}{x})}{\sqrt{x^2(x-\frac{27}{x^2})}}=\lim_{x\to\infty}\frac{x(2+\frac{3}{x})}{x\sqrt{(x-\frac{27}{x^2})}}=\lim_{x\to\infty}\frac{(2+\frac{3}{x})}{\sqrt{(x-\frac{27}{x^2})}}=\frac{\lim_{x\to\infty}(2)+\lim_{x\to\infty}(\frac{3}{x})}{\sqrt{\lim_{x\to\infty}(x)-\lim_{x\to\infty}(\frac{27}{x^2})}}=\frac{2+0}{\sqrt{\lim_{x\to\infty}(x)-0}}=0$$

12. Correct

$$\lim_{x\to\infty}\frac{x^2+2x-45}{4x^2-3x^2+7x-3}=\lim_{x\to\infty}\frac{x^2(1+\frac{2}{x}-\frac{45}{x^2})}{x^2(4-3+\frac{7}{x}-\frac{3}{x^2})}=\lim_{x\to\infty}\frac{(1+\frac{2}{x}-\frac{45}{x^2})}{(4+\frac{7}{x}-\frac{3}{x^2})}=\frac{\lim_{x\to\infty}(1)+\lim_{x\to\infty}(\frac{2}{x})-\lim_{x\to\infty}(\frac{45}{x^2})}{\lim_{x\to\infty}(4)+\lim_{x\to\infty}(\frac{7}{x})-\lim_{x\to\infty}(\frac{3}{x^2})}=\frac{1+0-0}{4+0-0}=\frac{1}{4}$$

Chapter 2. G. Limits-Continuity

We know that for a function f(x) to be continuous in x=a:

- f(a) has to be defined
- $\lim_{x\to a}f(x)$ exists
- $\lim_{x\to a}f(x)=f(a)$

The following function is represented below: $f(x) = \begin{cases} 2x-1, x \geq 2 \\ 3x-3, 1.5 < x < 2 \\ 0.5x, x \leq 1.5 \end{cases}$

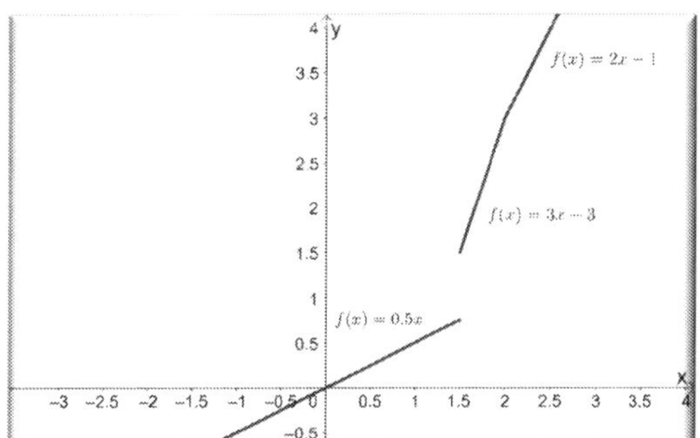

1. Correct

$\lim_{x\to 4} f(x) = 2(4) - 1 = 7$

$\lim_{x\to 4^-}[f(x)] = \lim_{x\to 4^+}[f(x)] = 7$

The function f(x) is continuous in x=4

2. Incorrect

$\lim_{x\to 2^-}(3x-3) = 3(2) - 3 = 3$

$\lim_{x\to 2^+}(2x-1) = 2(2) - 1 = 3$

So $\lim_{x\to 2^-}[f(x)] = \lim_{x\to 2^+}[f(x)] = f(2)$

The function f(x) is continuous in x=2

3. Correct

$\lim_{x\to 2} f(x) = 2(2) - 1 = 3$

$\lim_{x\to 2^-}[f(x)] = 3(2) - 3 = 3$

$\lim_{x\to 2^+}[f(x)] = 2(2) - 1 = 3$

$\lim_{x\to 2^-}[f(x)] = \lim_{x\to 2^+}[f(x)] = 3$

The function f(x) is continuous in x=1

4. Correct

The function f(x) is continuous in x=-7

5. Incorrect

$\lim_{x\to 1.5^-} f(x) = \lim_{x\to 1.5^-}[0.5(1.5)] = 0.75$

$\lim_{x\to 1.5^+} f(x) = \lim_{x\to 1.5^-}[3(1.5) - 3] = 1.3$

The limit for x=1.5 does not exist.

The function f(x) is not continuous in x=1.5

The function g(x) is defined as follows: $g(x) = \begin{cases} \frac{9x^2-1}{3x-1} \; for \; x \neq 1/3 \\ 0 \; for \; x = 1/3 \end{cases}$

6. Correct

The function g(x) exists for x=1/3. For x=1/3 g(x)=0

7. Incorrect

$\lim_{x \to 1/3^-} g(x) = \lim_{x \to 1/3^+} g(x) = \frac{9(\frac{1}{3})^2 - 1}{3(\frac{1}{3}) - 2} = \frac{1-1}{1-2} = 0 = g(\frac{1}{3})$

The limit $\lim_{x \to 1/3} g(x)$ does exist for x=1/3

8. Incorrect

The function g(x) is not continuous for x=0.33

- g(0.66) is not defined
- $\lim_{x \to 0.66^-} g(x) \frac{9(0.66)^2 - 1}{3(0.66) - 2} = \frac{2.96}{-0.02} = -\infty \neq \lim_{x \to 0.66^+} g(x) = +\infty$
- There is a vertical asymptote at x=0.66

The graph of g(x) is represented below.

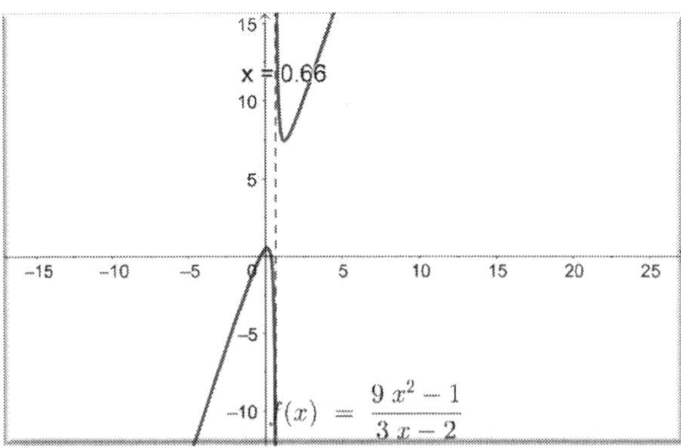

9. Correct

The function $f(x) = \begin{cases} 3x \text{ if } -1 \le x \le 2 \\ x + 2 \text{ if } x > 2 \end{cases}$ is continuous over interval [-1,2]

A linear function f(x)=3x is continuous for all real values of x. It is continuous on interval [-1,2]. The left-hand limit at 2 is 6 ; it exist and equals the value of the function at x=2.

10. Incorrect

The function $f(x) = \begin{cases} 5x \text{ if } -1 \le x \le 3 \\ 2x + 2 \text{ if } x > 3 \end{cases}$ is not continuous over [-5,5]

$\lim_{x \to 3^-} f(x) = 5(3) = 15; \lim_{x \to 3^+} f(x) = 2(3) + 2 = 8$

$\lim_{x \to 3} f(x)$ does not exist

So, f(x) is not continuous over [-5,5].

CHAPTER 3

Differentiation

"Take what you need, do what you should,
You will get what you want."

Gottfried Leibniz (1646-1716)

(conceived the ideas of differential and integral calculus)

Questions to be answered:

- One out of 4 cars in the world come from......
- Many rockets are prone to severe longitudinal vibrations like a stick toy from the 1960s, where the children would bounce along it, helped by a large spring. The Apollo astronauts felt very uncomfortable because of these longitudinal vibrations. What was the name of the toy where the name of the astronaut's sensations came from?
- A black bear's coat can be blue-gray or blue-black, brown and even sometimes...
- This Gaius had beaten the Germanic tribes in 102, and 100 BC. For these huge victories he was praised as "the third founder of Rome". He was elected consul of Rome for an unprecedented 7 times. What was his family name?
- The Saturn V was built as a three-stage vehicle. When one stage took over from the other, it was very complex and carefully controlled by the unit's computers and sequencers. This moment was called.....
- These mountains are the longest continental mountain range in the world. They form a "wall" of rock at the western edge of South America. What is the name of these mountains?
- Elk have feet like...
- Most of the lunar orbits of the Apollo tandem of command and service module (CSM) were ellipses. The point of the ellipse at which a spacecraft in lunar orbit is closest to the moon is called......
- Thirty five percent of the world's population drive on this side of the road. Which side is it?
- Each year, during the festival of Liberalia, in ancient Rome, the freeborn citizens boys from 14 to 17 years old marked the passage from childhood to manhood. What was the month this event took place?
- This glass pyramid was completed in 1989. It became a landmark of Paris "the city of lights". The pyramid is situated in the courtyard of the former royal palace now a famous museum called...

Interesting historical facts about Calculus

It is generally accepted that two men founded Calculus; Isaak Newton and Gottfried Leibniz.

Although Newton and Leibniz both had fundamental contributions in the creation of Calculus, their approach was quite different. (9)

Newton took into account that variables are changing with time, (9) whereas

Leibniz considered the variables x and y as extending over a number of infinitely close values. He was the first to introduce the notations dx and dy as differences between two successive values of these sequences.

Newton instead used the quantities x' and y', as notations for derivatives, which in his case were finite velocities, used to compute the tangent.

It is interesting to note that neither Leibniz nor Newton worked the concepts in terms of functions, but always in terms of graphs.

For Newton, the calculus was geometrical, while Leibniz took it towards analysis.

Leibniz was very aware of the importance of having a good notation while explaining the new concepts. He was very careful regarding the symbols he used.

Newton, instead, tended to use whatever notation he liked on that particular day.

Leibniz's notation became more used for generalizing calculus to multiple variables.

Leibniz's notation emphasized the operator aspect of the derivative as well as the integral. Much of the notation that we are using in Calculus today is what Leibniz used.

3.B. Differentiation

Definition of derivatives

Theory and Examples

We know from Pre-Calculus that the slope of a straight line that passes through points $A(x_1, y_1)$ and $B(x_2, y_2)$ can be calculated as:

$slope = \frac{y_2 - y_1}{x_2 - x_1}$

EXAMPLE

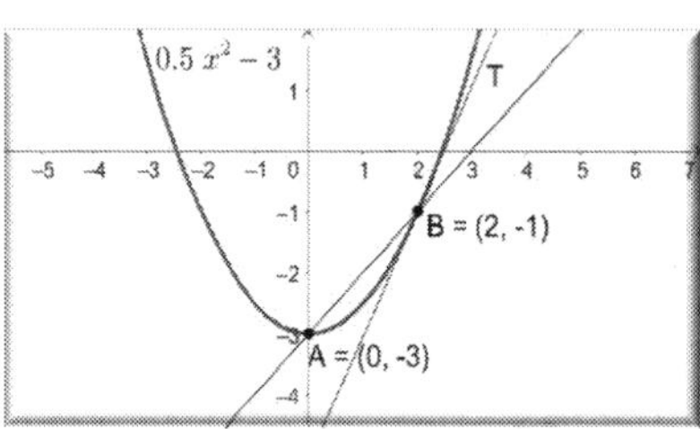

As we can see here, the slope of the line that passes through A(0,-3) and B(2,-1) is:

$slope = \frac{y_2 - y_1}{x_2 - x_1} = \frac{-1-(-3)}{2-0} = \frac{-1+3}{2} = \frac{2}{2} = 1$

The slope of the line that only passes through point B and is the tangent to the graph at this point, is defined by the limit of the slope relation when x is approaching 2.

This limit of the slope when x approaches zero is called derivative.

The formula of the derivative is: $f'(x) = \lim_{h \to 0} \frac{y_2 - y_1}{x_2 - x_1} = \lim_{h \to 0} \frac{f(x+h) - f(x)}{h}, where\ h\ is\ a\ very\ small\ number.$

First, we have to calculate the limit of the expression for the difference between x_1 and $x_2 = x_1 + h$, where h is an extremely small value approaching zero.

So, $x_2 - x_1 = x_1 + h - x_1 = h$

The derivative of $y = f(x) = 0.5x^2 - 3\ is\ f' = \lim_{h \to 0} \frac{f(x+h) - f(x)}{h} = \lim_{h \to 0} \frac{0.5(x+h)^2 - 3 - [0.5x^2 - 3]}{h} =$

$\lim_{h \to 0} \frac{0.5(x^2 + 2xh + h^2) - 3 - 0.5x^2 + 3}{h} = \lim_{h \to 0} \frac{0.5x^2 + xh + 0.5h^2 - 3 - 0.5x^2 + 3}{h} = \lim_{h \to 0} \frac{xh + 0.5h^2}{h} = \lim_{h \to 0} \frac{h(x + 0.5h)}{h} = \lim_{h \to 0}(x + 0.5h) =$

$\lim_{h \to 0}(x) + \lim_{h \to 0}(0.5h) = x$

For x=2 $\lim_{h \to 0} \frac{f(x+h) - f(x)}{h} = 2$

EXAMPLE

If we consider the very small increase from x as Δx, where $\Delta x \to 0$, then the limit of:

$\frac{f(x+h) - f(x)}{h}\ of\ f(x) = 0.5x^2 - 3$ can be written:

$f' = \lim_{\Delta x \to 0} \frac{f(x + \Delta x) - f(x)}{\Delta x} = \lim_{\Delta x \to 0} \frac{0.5(x + \Delta x)^2 - 3 - [0.5x^2 - 3]}{\Delta x} =$

$\lim_{\Delta x \to 0} \frac{0.5(x^2 + 2x\Delta x + \Delta x^2) - 3 - 0.5x^2 + 3}{\Delta x} = \lim_{\Delta x \to 0} \frac{0.5x^2 + x\Delta x + 0.5\Delta x^2 - 3 - 0.5x^2 + 3}{\Delta x} = \lim_{\Delta x \to 0} \frac{x\Delta x + 0.5\Delta x^2}{\Delta x} = \lim_{\Delta x \to 0} \frac{\Delta x(x + 0.5\Delta x)}{\Delta x} =$

$\lim_{\Delta x \to 0}(x + 0.5\Delta x) = \lim_{\Delta x \to 0}(x) + \lim_{\Delta x \to 0}(0.5\Delta x) = x$

Chapter 3. B. Differentiation-Definition of derivatives

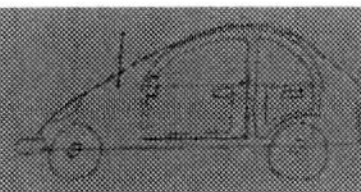

One out of 4 cars in the world come from……

Determine which answer is correct. In the table at the bottom of the page cross off all the letters for the correct answers.

The formula of the derivative is: $f'(x) = \lim\limits_{h \to 0} \frac{f(x+h)-f(x)}{h}$

Using this formula, calculate the following derivatives.

1)$f(x) = 2x + 45$ so $f'(x) = 2$

2) $f(x) = 3x^2 + 5x - 4$ so $f'(x) = 6x - 5$

3) $f(x) = Ax^2 + Bx - 37$ so $f'(x) = 2Ax + B$

Using Δx notation instead of h, find the derivative of the following functions:

4) $f(x) = 5x^2 - 4x + 3$ so $f'(x) = 10x + 3$

5) $f(x) = 7x - 27$ so $f'(x) = 7$

6) $f(x) = x^{-1}$ so $f'(x) = x$

7) $f(x) = 3x^{-1} + x$ so $f'(x)$ =6x

Using the formula $f'(x) = \lim\limits_{h \to 0} \frac{f(x+h)-f(x)}{h}$ find the following derivatives.

8) $f(x) = x^{-1} + x^2$ so $f'(x) = -x^{-2} + 2x = 2x - x^{-2}$

9) $f(x) = x^2 + 3x$ $sofor$ $x = 2$, $f'(2) = 10$

10) $f(x) = x^2 + 33$ $sofor$ $x = 3$, $f'(3) = 6$

1	2	3	4	5	6	7	8	9	10
E	C	O	H	U	I	N	M	A	K

Chapter 3. C. Differentiation-Notation

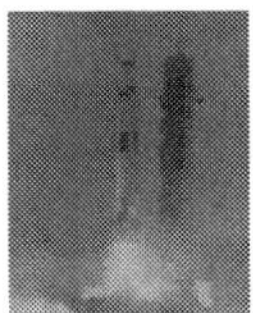

Many rockets are prone to severe longitudinal vibrations like a stick toy in the 1960s, where the children would bounce along it, helped by a large spring. The Apollo astronauts felt very uncomfortable because of these longitudinal vibrations. What was the name of the toy where the name of the astronaut's sensations came from?

Determine which answer is correct. In the table at the bottom of the page cross off all the letters for the correct answers.

1) Gottfried Leibniz's notation is: $\frac{dy}{dx}$

2) Another notation is: $f(x) = y$

3) Joseph Louis Lagrange's notation is $f'(x)$

4) An interesting notation is d=V*t

5) Leonhard Euler's notation for the second derivative is: D_x^2

6) Travis formula is F=k/e

7) One of Isaac Newton's notation is: $\dot{y}$ for first derivative

8) Henry's formula is: SO=CCER

9) Another formula used by Isaac Newton was: $\dot{x}$

10) Isaac Newton also used this notation for first derivative: □$\acute{y}$

1	2	3	4	5	6	7	8	9	10
A	P	R	O	L	G	I	O	L	M

3.D. Differentiation

a Rate of change - Average versus Instantaneous

Theory and Examples

The **rate of change** tells us how fast y is changing as x is changing.

EXAMPLE

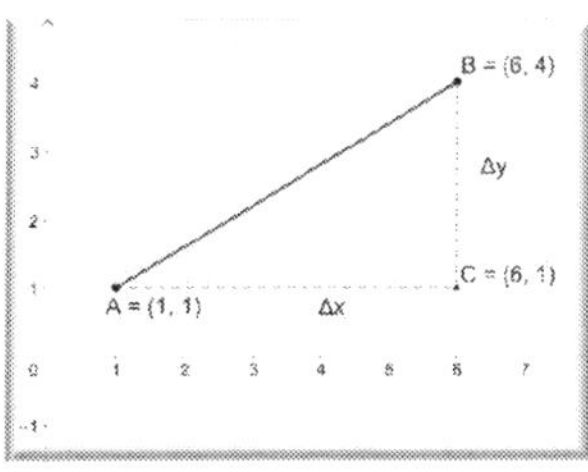

We want to find the rate of change between points A (1,1) and B (6,4). As we go from A to B x varies from 1 to 6, while y varies from 1 to 4.

The rate of change is then given by the ratio $\frac{\Delta y}{\Delta x}$

In this case $\Delta y = 4 - 1 = 3$

$\Delta x = 6 - 1 = 5$

So, $\frac{\Delta y}{\Delta x} = \frac{3}{5}$

When we have a function f(x), the rate of change between two points $D[x_1, f(x_1)]$ and $E[x_2, f(x_2)]$ that belong to the graph of the function, then the average rate of change between D and E is given by: $\frac{\Delta y}{\Delta x} = \frac{f(x_2)-f(x_1)}{x_2-x_1}$

EXAMPLE

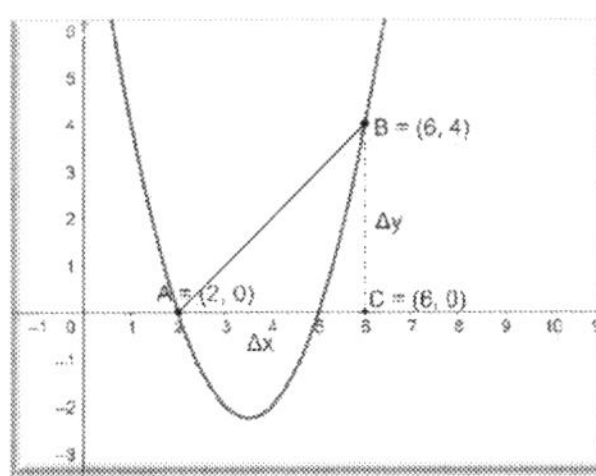

Here, $x_1 = 2$ *and* $x_2 = 6$, so, $f(x_1) = 0$ *and* $f(x_2) = 4$

The average rate of change is:

$\frac{\Delta y}{\Delta x} = \frac{f(x_2)-f(x_1)}{x_2-x_1} = \frac{4-0}{6-2} = \frac{4}{4} = 1$

The Instantaneous rate of change happens when Δx is extremely small, approaches zero.

The instantaneous rate of change is given by the limit when Δx is extremely small, approaches zero.

EXAMPLE

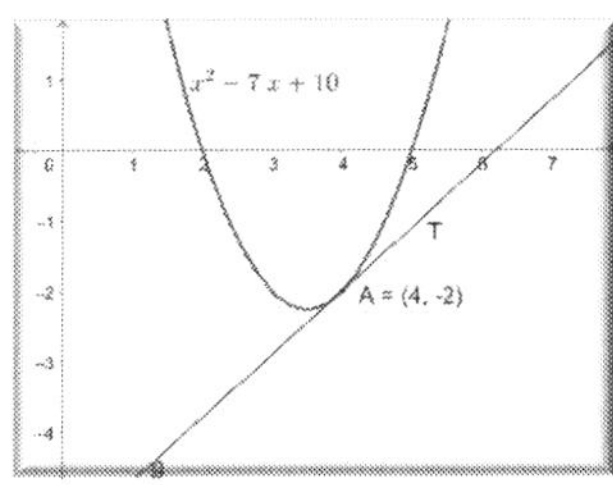

Point B, extremely close to A, has the coordinates $[x + \Delta x, f(x + \Delta x)]$ The formula we use is:

$$f'(x) = \lim_{\Delta x \to 0} \frac{f(x+\Delta x)-f(x)}{\Delta x} = \lim_{\Delta x \to 0} \frac{(x+\Delta x)^2-7(x+\Delta x)+10-(x^2-7x+10)}{\Delta x} =$$

$$\lim_{\Delta x \to 0} \frac{x^2+2x\Delta x+(\Delta x)^2-7x-7\Delta x+10-x^2+7x-10}{\Delta x} = \lim_{\Delta x \to 0} \frac{2x\Delta x+(\Delta x)^2-7\Delta x}{\Delta x} =$$

$$\lim_{\Delta x \to 0} \frac{\Delta x(2x-\Delta x-7)}{\Delta x} = \lim_{\Delta x \to 0} (2x - \Delta x - 7) = 2x - 7$$

So, $f'(x) = 2x - 7$

For x=4 the instantaneous rate of change, called derivative is $f'(4) = 2(4) - 7 = 8 - 7 = 1$

Chapter 3. D. a. Rate of change- Average versus Instantaneous

A black bear's coat can be blue-gray or blue-black, brown and even sometimes...

Determine which answer is correct. In the table at the bottom of the page cross off all the letters for the correct answers.

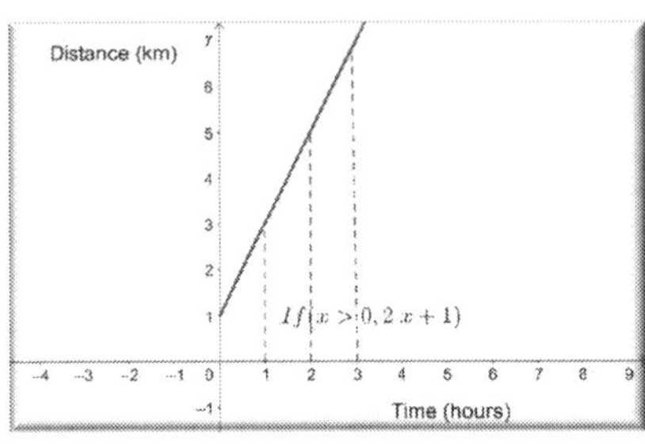

1) The average rate of change (speed) between 1 and 3 hours is: 2km/h.

2) The average rate of change (speed) between 0 and 3 hours is: 3km/h.

A tire rolls by the relation between distance and time as follows: $s(t) = 2t^2 + 4t + 3$

3) The instantaneous velocity (first derivate) at t=2 seconds is: 12m/s.

4) The instantaneous velocity (first derivate) at t=5 seconds is: 15m/s.

If $R(t) = \frac{3000t^2+500t}{5} + 20{,}000$ represent the revenue a company earns in time.

5) After 10 days, the revenue is increasing with a speed of $32,100 per day.

6) After 30 days, the revenue is increasing with a speed of $45,000 per day.

The following questions follow the graph below. The graph represents the function: $f(x) = -0.2x^2 + 2x + 1$

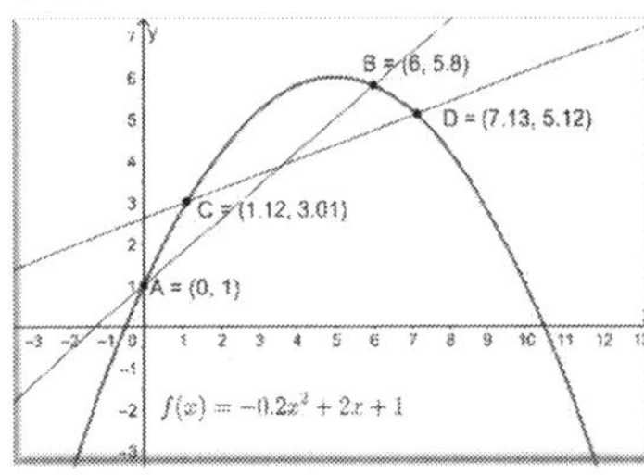

7) The average rate of change between point A and B is $\frac{4.8}{6}$

8) The average rate of change between point C and D is $\frac{6}{5}$

9) The instantaneous rate of change (first derivate) at point C equals 1.552

10) The instantaneous rate of change (first derivate) at point D equals 6.1

1	2	3	4	5	6	7	8	9	10
A	W	O	H	F	I	L	T	U	E

3.D. Differentiation

b Rate of change - Slope of secant and tangent lines

Theory and Examples

A secant is a straight line that passes through a curve and cuts it into two or more parts.

EXAMPLE

The secant DA intersects the graph in points D and A. To be able to calculate the slope, we create our own right-angle triangle ΔABC where angle $\measuredangle CBA$ is 90^0. The slope of the secant DA is calculated using the points C (point 1), and A (point 2), by using the slope relation:

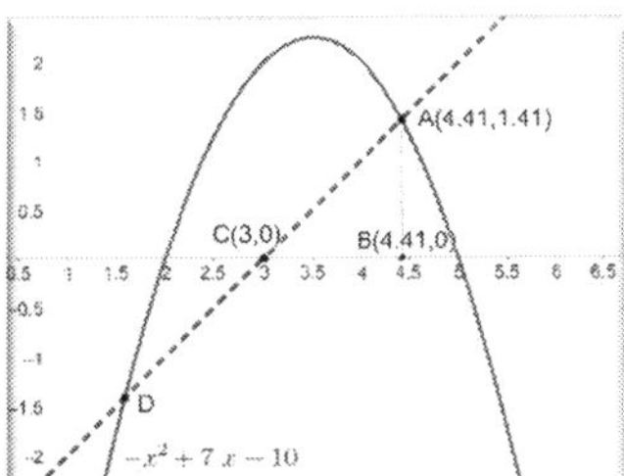

$$slope = \frac{\Delta y}{\Delta x} = \frac{y_2-y_1}{x_2-x_1} = \frac{1.41-0}{4.41-3} = \frac{1.41}{1.41} = 1$$

A **tangent** to a graph is when the line touches the graph in only one point.
In this case, the distance on x axis is between the x coordinates is x+Δx-x=Δx, and on y axis is y+$\Delta y - y$=$f(x+\Delta x) - f(x) = \Delta y$. We consider Δx extremely small, approaching zero.

EXAMPLE

To be able to calculate the slope of a tangent, we consider the limit of the ratio $\frac{\Delta y}{\Delta x}$ for Δx approaching zero. We know that the function that represents the graph is given by: $f(x) = -x^2 + 7x - 10$

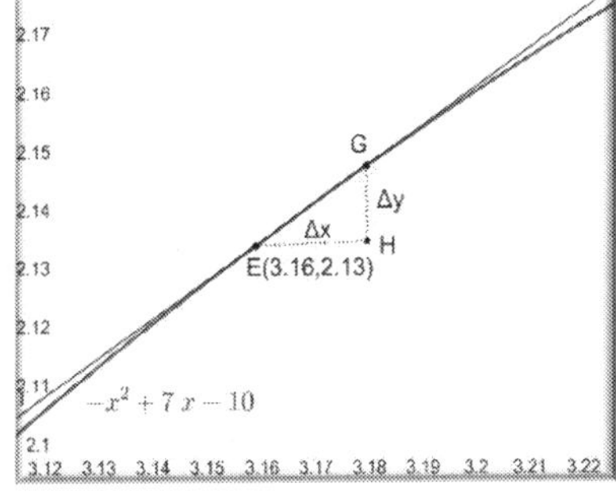

So:

$$slope = \lim_{\Delta x \to 0} \frac{\Delta y}{\Delta x} =$$

$$\lim_{\Delta x \to 0} \frac{f(x+\Delta x)-f(x)}{\Delta x} = \lim_{\Delta x \to 0} \frac{-(x+\Delta x)^2+7(x+\Delta x)-10-(-x^2+7x-10)}{\Delta x} =$$

$$\lim_{\Delta x \to 0} \frac{-(x^2+2x\Delta x+\Delta x^2)+7(x+\Delta x)-10-(-x^2+7x-10)}{\Delta x} =$$

$$\lim_{\Delta x \to 0} \frac{-x^2-2x\Delta x-\Delta x^2+7x+7\Delta x-10+x^2-7x+10}{\Delta x} =$$

$$\lim_{\Delta x \to 0} \frac{-2x\Delta x-\Delta x^2+7\Delta x}{\Delta x} = \lim_{\Delta x \to 0} \frac{\Delta x(-2x-\Delta x+7)}{\Delta x} = \lim_{\Delta x \to 0}(-2x+7) = -2x+7$$

The slope of the tangent at the point E(3.16,2.13) is given by:

$slope = -2(3.16) + 7 = -6.32 + 7 = 0.68$

Chapter 3. D. b. Rate of change- Slope of secant and tangent lines

This Gaius had beaten the Germanic tribes in 102, and 100 BC. For these huge victories he was praised as "the third founder of Rome". He was elected consul of Rome for an unprecedented 7 times. What was his family name?

Determine which answer is correct. In the table at the bottom of the page cross off all the letters of the correct answers.

The next questions follow the graph below. The function represented below is:

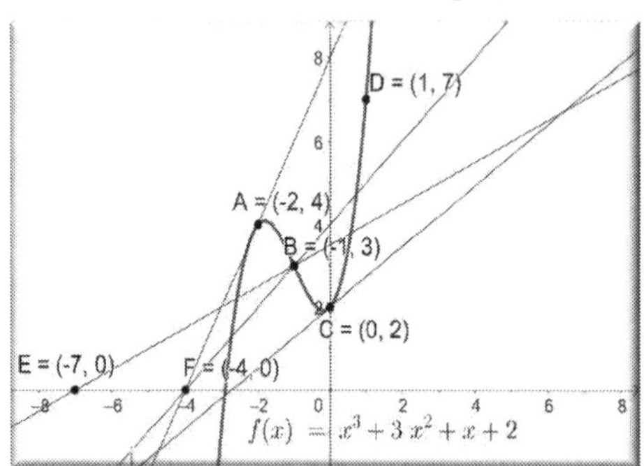

$f(x) = x^3 + 3x^2 + x + 2$

1) The slope of the secant that passes through the points E and B is: $\frac{1}{2}$

2) The slope of the secant that passes through the points F and B is: $\frac{2}{3}$

3) The slope of the tangent to the graph in point A equals 1

4) The slope of the tangent to the graph in point C equals 2

5) The slope of the tangent to the graph in point D equals 8

6) The slope of the tangent to the graph in point B equals -2

7) The slope of the secant that passes through the points A and B is: $\frac{2}{5}$

8) The slope of the secant that passes through the points B and C is: -1

9) The slope of the secant that passes through the points B and D is: $-\frac{2}{7}$

10) The slope of the secant that passes through the points C and D is: $\frac{1}{5}$

1	2	3	4	5	6	7	8	9	10
O	M	L	A	R	E	I	C	U	S

3.E. Transcendental Functions – Logarithmic, Exponential, Trigonometric

Theory and Examples

Formulas used for differentiation of transcendental functions.

a) $f(x) = ln|x|\ then\ f'(x) = \frac{1}{x}$

EXAMPLE

If $f(x) = 2ln|x| + 3\ then\ f'(x) = \frac{2}{x}$

b) $f(x) = e^x\ then\ f'(x) = e^x \quad f(x) = a^x\ then\ f'(x) = a^x * \ln(a)$

EXAMPLE

If $f(x) = 3^x - 4x + 5\ \ then\ f'(x) = 3^x * \ln(3) - 4$

c) $f(x) = \sin(x)\ then\ f'(x) = \cos(x) \quad f(x) = \cos(x)\ then\ f'(x) = -sin(x)$

EXAMPLE

If $f(x) = 3\text{x} - \sin(x)\ then\ f'(x) = 3 - \cos(x)$

d) $f(x) = \tan(x)\ then\ f'(x) = sec^2(x) \quad f(x) = \cot(x)\ then\ f'(x) = -csc^2(x)$

EXAMPLE

If $f(x) = \tan(x) + 5\ then\ f'(x) = sec^2(x)$

e) $f(x) = \sec(x)\ then\ f'(x) = \sec(x) * \tan(x) \quad f(x) = \csc(x)\ then\ f'(x) = -\csc(x) * \cot(x)$

EXAMPLE

If $f(x) = 2\sec(x) + \tan(x)\ then\ f'(x) = 2\sec(x) * \tan(x) + sec^2(x)$

f) $f(x) = sin^{-1}(x)\ then\ f'(x) = \frac{1}{\sqrt{1-x^2}} \quad f(x) = cos^{-1}(x)\ then\ f'(x) = \frac{-1}{\sqrt{1-x^2}}$

EXAMPLE

If $f(x) = 3x - sin^{-1}(x)\ then\ f'(x) = 3 - \frac{1}{\sqrt{1-x^2}} = \frac{3\sqrt{1-x^2}-1}{\sqrt{1-x^2}}$

g) $f(x) = tan^{-1}(x) = \frac{1}{x^2+1} \quad f(x) = cot^{-1}(x) = \frac{-1}{x^2+1}$

EXAMPLE

If $f(x) = 7x + tan^{-1}(x)\ then\ f'(x) = 7 + \frac{1}{x^2+1} = \frac{7(x^2+1)+1}{x^2+1} = \frac{7x^2+8}{x^2+1}$

Chapter 3. E. Transcendental Functions – Logarithmic, Exponential, Trigonometric

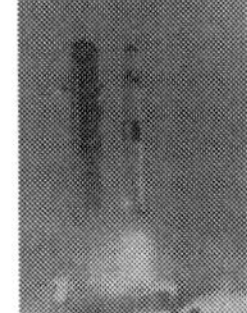

The Saturn V was built as a three-stage vehicle. When one stage took over from the other it was very complex and carefully controlled by the unit's computers and sequencers. This moment it was called.....

Determine which answer is correct. In the table at the bottom of the page cross off all the letters of the correct answers.

1) If $f(x) = 2e^x - 7\ln(x)\,;\ then\ f'(x) = 6x - 7$

2) If $f(x) = 3\ln(x) + 5e^x - ex;\ then\ f'(x) = \frac{3}{x} + 5e^x - e$

3) If $f(x) = 37 - e^x;\ then\ f'(x) = 2x + e^x$

4) If $f(x) = 73 + \ln(124) - e^2;\ then\ f'(x) = -2e$

5) If $f(x) = -4e^x + \ln(27)\,;\ then\ f'(x) = -4e^x$

6) If $f(x) = \cos(x) + 3x;\ then\ f'(x) = \tan(x) + 3$

7) If $(x) = 2\ln(x) + 3e^x + \tan(x)\,;\ then\ f'(x) = 2x + 3e^x + sec^2(x)$

8) If $(x) = \frac{3}{7}\ln(x) + \frac{4}{9}e^x + \frac{5}{9}\tan(x)\,;\ then\ f'(x) = \frac{3}{7}\mathrm{x} + \frac{4}{9}e^x + \frac{5}{9}sec^2(x)$

9) If $(x) = \frac{5}{9}\ln(2) + \frac{1}{2}e^x + \cos(x)\,;\ then\ f'(x) = \frac{1}{2}e^x - \sin(x)$

10) If $(x) = 4e^x + \frac{5}{6}\ln(x) - 6\cot(x)\,;\ then\ f'(x) = 4e^x + \frac{5}{6x} + 6cos^2(x)$

1	2	3	4	5	6	7	8	9	10
S	E	T	A	R	G	I	N	O	G

3.F. Differentiation

a. Differential rules - Power

Theory and Examples

When we have a polynomial function like $f(x) = x^n$ where n is a positive integer, the derivative of this function is:
$f'(x) = nx^{n-1}$

EXAMPLE

If $f(x) = x^3$ $then, f'(x) = 3x^{3-1} = 3x^2$
If $f(x) = 2x^4 - 3x^3 + 1$ $then, f'(x) = 2(4)x^{4-1} - 3(3)x^{3-1} + 0 = 8x^3 - 9x^2$

When we have a function where the exponent is a rational number like $f(x) = x^{\frac{m}{n}}$, m,n $\in Z \neq 0$, the derivative of this function is:
$f'(x) = \frac{m}{n}x^{\frac{m}{n}-1} = \frac{m}{n}x^{\frac{m-n}{n}}$

EXAMPLE

If $f(x) = 3x^{\frac{1}{2}}$ $then, f'(x) = \frac{3}{2}x^{\frac{1}{2}-1} = \frac{3}{2}x^{-\frac{1}{2}} = \frac{3}{2} * \frac{1}{x^{\frac{1}{2}}} = \frac{3}{2\sqrt{x}} = \frac{3}{2\sqrt{x}}\frac{\sqrt{x}}{\sqrt{x}} = \frac{3\sqrt{x}}{2x}$

When the unknown is at the denominator.

EXAMPLE

If $f(x) = \frac{3}{x}$ $; x \neq 0$ $then$ $f'(x) = 3x^{-1} = (-1)(3)x^{-1-1} = -3x^{-2} = -\frac{3}{x^2}$

In case we have a radical function.

EXAMPLE

If $f(x) = 7\sqrt{x} = 7x^{\frac{1}{2}}; x \geq 0$ $so, f'(x) = \frac{7}{2}x^{\frac{1}{2}-1} = \frac{7}{2}x^{-\frac{1}{2}} = \frac{7}{2\sqrt{x}} = \frac{7\sqrt{x}}{2x}$

If $f(x) = \frac{5}{\sqrt{x}} = 5x^{-\frac{1}{2}}; x > 0$

$so, f'(x) = -\frac{5}{2} * x^{-\frac{1}{2}-1} = -\frac{5}{2}x^{\frac{-1-2}{2}} = -\frac{5}{2}x^{\frac{-3}{2}} = -\frac{5}{2x^{\frac{3}{2}}} = -\frac{5}{2\sqrt{x^3}} = -\frac{5\sqrt{x^3}}{2x^3} = -\frac{5x\sqrt{x}}{2x^3} = -\frac{5\sqrt{x}}{2x^2}$

Chapter 3. F. a. Differentiation rules - Power

These mountains are the longest continental mountain range in the world. They form a nonstop "wall" of rock at the western edge of South America. What is the name of these mountains?

Determine which answer is correct. In the table at the bottom of the page cross off all the letters of the correct answers.

1) If $f(x) = 3x^2 - 7x + 89 \; then \; f'(x) = 6x - 7$

2) If $f(x) = \frac{2}{5}x^2 + 27x - 63 \; then \; f'(x) = 4x + 27$

3) If $f(x) = \frac{3}{7}x^4 - 2x^3 - 6x^2 + 81 \; then \; f'(x) = 12x^3 + 6x^2 + 81$

4) If $f(x) = x^4 - 2x^3 - 6x^2 + 81 \; then \; f'(x) = 4x^3 - 6x^2 - 12x$

5) If $f(x) = \frac{1}{x} \; then \; f'(x) = x^2$

6) If $f(x) = \frac{1}{x^4} \; then \; f'(x) = -4x^{-5}$

7) If $f(x) = \sqrt{x^3} \; then \; f'(x) = -3x^{-2}$

8) If $f(x) = \frac{\sqrt{x}}{x^4} \; then \; f'(x) = 3x^{\frac{3}{5}}$

9) If $f(x) = \frac{\sqrt{x}}{x^2} + 3x \; then \; f'(x) = -\frac{3}{2x\sqrt{x}} + 3$

10) If $f(x) = \frac{x^3}{x^{-2}} \; then \; f'(x) = 5x^4$

1	2	3	4	5
C	A	N	O	D
6	7	8	9	10
R	E	S	U	N

3.F Differentiation

b. Differential rules - Product

Theory and Examples

If we have two functions f and g, the derivative of the product between f and g is:
$(f * g)' = f'g + fg'$

EXAMPLE

If $f(x) = 2x - 3$ and $g(x) = 5x^3 + 3x^2$
$[f(x) * g(x)]' = [(2x-3) * (5x^3 + 3x^2)]' = (2x-3)'(5x^3 + 3x^2) + (2x-3)(5x^3 + 3x^2)' =$
$2(5x^3 + 3x^2) + (2x-3)(15x^2 + 6x) = 10x^3 + 6x^2 + 30x^3 + 12x^2 - 45x^2 - 18x = 40x^3 - 27x^2 - 18x$

EXAMPLE

If $f(x) = \sqrt{x} - 3$ and $g(x) = 3x^2 + 7x$
$[f(x) * g(x)]' = [(\sqrt{x} - 3) * (3x^2 + 7x)]' = (\sqrt{x} - 3)'(3x^2 + 7x) + (\sqrt{x} - 3)(3x^2 + 7x)' =$
$(x^{\frac{1}{2}})'(3x^2 + 7x) + (\sqrt{x} - 3)(6x + 7) = \left(\frac{1}{2}x^{-\frac{1}{2}}\right)(3x^2 + 7x) + 6x\sqrt{x} + 7\sqrt{x} - 18x - 21 = \frac{3}{2}x^{\frac{3}{2}} + \frac{7}{2}\sqrt{x} +$
$6x\sqrt{x} + 7\sqrt{x} - 18x - 21 = \frac{3}{2}x\sqrt{x} + \frac{7}{2}\sqrt{x} + 6x\sqrt{x} + 7\sqrt{x} - 18x - 21 = \frac{15}{2}x\sqrt{x} + \frac{21}{2}\sqrt{x} - 18x - 21$

EXAMPLE

If $f(x) = ln|x|$ and $g(x) = x^3$
$[f(x) * g(x)]' = [ln|x| * x^3]' = (ln|x|)'(x^3) + ln|x|(x^3)' = \frac{x^3}{x} + ln|x| * 3x^2 = x^2 + 3x^2 ln|x| = x^2(1 +$
$3ln|x|)$

Chapter 3. F. b. Differentiation rules – Product

Elk have feet like ……

Determine which answer is correct. In the table at the bottom of the page cross off all the letters of the correct answers.

1) If $f(x) = 2x^2\cos(x)\,; then\ f'(x) = 4xcos(x) - 2x^2\sin(x)$

2) If $f(x) = 3x^4\sqrt{x}; then\ f'(x) = 12x + 3\sqrt{x}$

3) If $f(x) = 4x^2\tan(x)\,; then\ f'(x) = (8x)\tan(x) + 4x^2 sec^2(x)$

4) If $f(x) = \frac{2}{5}x^2e^x; then\ f'(x) = 4x + e^x$

5) If $f(x) = 2^x\sin(x)\,; then\ f'(x) = 2^x\ln 2\sin(x) + 2^x\cos(x)$

6) If $f(x) = 3\ln|x|x^2; then\ f'(x) = 6x$

7) If $f(x) = \ln|x|\sqrt{x}; then\ f'(x) = x^{-\frac{1}{2}}(1 + \frac{1}{2}\ln|x|)$

8) If $f(x) = \cos(x) * e^x; then\ f'(x) = e^x\cos(x)$

9) If $f(x) = e^x3^x; then\ f'(x) = e^x3^x[1 + \ln(3)]$

10) If $f(x) = e^xx^5; then\ f'(x) = e^xx^4(x + 5)$

1	2	3	4	5
A	C	R	O	L
6	7	8	9	10
W	U	S	I	N

3.F. Differentiation

c. Differential rules - Quotient

Theory and Examples

If we want to find the derivative of a quotient of f(x) and g(x), we apply this formula:

$[\frac{f(x)}{g(x)}]' = \frac{[f(x)]'g(x)-f(x)[g(x)]'}{[g(x)]^2}$

EXAMPLE

$f(x) = x^2 + 1, and\ g(x) = 2x$

$[\frac{f(x)}{g(x)}]' = \frac{[f(x)]'g(x)-f(x)[g(x)]'}{[g(x)]^2} = \frac{(x^2+1)'(2x)-(x^2+1)(2x)'}{4x^2} = \frac{(2x)(2x)-2(x^2+1)}{4x^2} = \frac{4x^2-2x^2-2}{4x^2} = \frac{2(x^2-1)}{4x^2} = \frac{x^2-1}{2x^2}$

EXAMPLE

If $f(x) = \frac{3x^2-4x+5}{6x-7}; then\ f'(x) = [\frac{s(x)}{m(x)}]' = \frac{[s(x)]'m(x)-s(x)[m(x)]'}{[m(x)]^2} = \frac{[3x^2-4x+5]'(6x-7)-(3x^2-4x+5)[(6x-7)]'}{[(6x-7)]^2} =$

$\frac{(6x-4)(6x-7)-(3x^2-4x+5)6}{(6x-7)^2} = \frac{36x^2-42x-24x+28-(18x^2-24x+30)}{(6x-7)^2} = \frac{36x^2-42x-24x+28-18x^2+24x-30}{(6x-7)^2} = \frac{18x^2-42x-2}{(6x-7)^2}$

EXAMPLE

If $f(x) = 2x^3 - 7, and\ g(x) = 2\sqrt{x}$

We can write $\sqrt{x}$ as $x^{\frac{1}{2}}\ so, (x^{\frac{1}{2}})' = \frac{1}{2}x^{\frac{1}{2}-1} = \frac{1}{2}x^{-\frac{1}{2}} = \frac{1}{2x^{\frac{1}{2}}} = \frac{1}{2\sqrt{x}}$

$[\frac{f(x)}{g(x)}]' = \frac{[f(x)]'g(x)-f(x)[g(x)]'}{[g(x)]^2} = \frac{(2x^3-7)'(2\sqrt{x})-(2x^3-7)(2\sqrt{x})'}{(2\sqrt{x})^2} = \frac{(6x^2)(2\sqrt{x})-(2x^3-7)\frac{2}{2\sqrt{x}}}{4x} = \frac{(6x^2)(2\sqrt{x})(\sqrt{x})-(2x^3-7)}{4x*\sqrt{x}} =$

$\frac{12x^3-2x^3+7}{4x\sqrt{x}} = \frac{10x^3+7}{4x\sqrt{x}}$

Chapter 3. F. c. Differentiation rules - Quotient

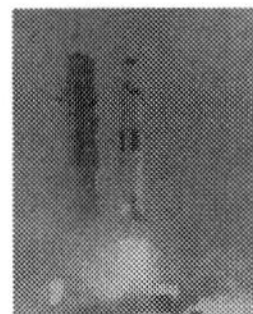

Most of the lunar orbits of the Apollo tandem of command and service module (CSM) were ellipses. The point of the ellipse at which a spacecraft in lunar orbit is closest to the moon is called……

Determine which answer is correct. In the table at the bottom of the page cross off all the letters of the correct answers.

1) If $f(x) = \tan(x)\,; then\ f'(x) = [\sec(x)]^2$

2) If $f(x) = \frac{\sin(x)}{x^2}; then\ f'(x) = [\csc(x)]^2$

3) If $f(x) = \frac{x}{e^x}; then\ f'(x) = \frac{x}{e^x}$

4) If $f(x) = \frac{\sin(x)}{x^2+1}; then\ f'(x) = \frac{\cos(x)}{[x^2+1]^2}$

5) If $f(x) = \frac{x^2+2x}{\sqrt{x}}; then\ f'(x) = 3x + 1$

6) If $f(x) = 2x\sec(x) = \frac{2x}{cox(x)}; then\ f'(x) = \frac{2[1-xtan(x)]}{\cos(x)}$

7) If $f(x) = \frac{3x^2-4x+2}{\sin(x)}; then\ f'(x) = \frac{6x-4}{[\sin(x)]^2}$

8) If $f(x) = \frac{x^2+2x-3}{5x+4}; then\ f'(x) = \frac{x^2+2x-3}{(5x+4)^2}$

9) If $f(x) = \frac{x^2}{lnx}; then\ f'(x) = 2x$

10) $f(x) = \frac{2x^2-4x+6}{2x}; then\ f'(x) = 4x^2 - 2x$

1	2	3	4	5	6	7	8	9	10
A	P	E	R	I	K	L	U	N	E

3.F. Differentiation

d. Differential rules - Chain

Theory and Examples

We have a function f(u), but u is as well a function of a variable x ; u(x). We will have the situation of a function of a function f(u(x)). To find the derivative of f(u(x)), we apply the "chain" rule:

$f'(u(x)) = f'(u)[u'(x)]$

The derivative of a function f of a function u equals the derivative of f(u) times the derivative of u(x).

EXAMPLE

If $f(x) = (6x + 7)^3, u(x) = 6x + 7, and\ f(u) = u^3$

Here we decide that u(x) is 6x+7 so, f(u) becomes $f(u) = u^3$

Applying the chain rule, we have:

$f'(x) = f'(u)[u'(x)] = 3(6x + 7)^2(6x + 7)' = 3(6x + 7)^2 * 6 = 18(6x + 7)^2$

EXAMPLE

If $f(x) = \sqrt{2x^2 + x}, u(x) = 2x^2 + x\ then, f(u) = \sqrt{u}$

We can write $\sqrt{u}$ as $u^{\frac{1}{2}}\ so, (u^{\frac{1}{2}})' = \frac{1}{2}u^{\frac{1}{2}-1} = \frac{1}{2}u^{-\frac{1}{2}} = \frac{1}{2u^{\frac{1}{2}}} = \frac{1}{2\sqrt{u}}$

Applying the chain rule, we have:

$f'(x) = f'(u)[u'(x)] = (\sqrt{u})'[u(x)]' = \frac{1}{2\sqrt{u}}(2x^2 + x\ )' = \frac{4x+1}{2\sqrt{2x^2+x}}$

EXAMPLE

If $f(x) = \ln(2x - 3)\,, u(x) = 2x - 3\ then, f(u) = \ln(u)$

We know that $[\ln(u)]' = \frac{1}{u}$

Applying the chain rule, we have:

$f'(x) = f'(u)[u'(x)] = \frac{1}{u}(2x - 3)' = \frac{2}{2x-3}$

Chapter 3. F. d. Differentiation rules – Chain

Thirty five percent of the world's population drive on this side of the road. Which side is it?

Determine which answer is correct. In the table at the bottom of the page cross off all the letters of the correct answers.

1) If $f(x) = (x^2 - 4x + 5)^2; then\ f'(x) = 2(x^2 - 4x + 5)(2x - 4)$

2) If $f(x) = (3x^2 + 7x)^{\frac{3}{2}}; then\ f'(x) = \frac{3}{2}\sqrt{3x^2 + 7x}\,(6x + 7)$

3) If $f(x) = \sqrt{x^3 - 5x}; then\ f'(x) = \frac{(3x^2-5)}{\sqrt{x^3-5x}}$

4) If $f(x) = \sqrt{e^x - 2x}; then\ f'(x) = \frac{(e^x-2)}{2\sqrt{(e^x-2x}}$

5) If $f(x) = \sqrt{\sin(x)}; then\ f'(x) = \frac{1}{2}\tan(x)$

6) If $f(x) = 2[\ln(4x - 3)]; then\ f'(x) = \frac{8}{4x-3}$

7) If $f(x) = [\ln(e^x - e^{-x})]; then\ f'(x) = \frac{e^x+e^{-x}}{2(e^x-e^{-x})}$

8) If $f(x) = \sin(x - 1)\ ;\ then\ f'(x) = \cos(x - 1)$

9) If $f(x) = e^x + 2\ln(x + 7)\ ;\ then\ f'(x) = e^x + \frac{x}{2(x+7)}$

10) $f(x) = \ln(x^2 - 4x)\ ;\ then\ f'(x) = \frac{2(x-2)}{x(x-4)}$

1	2	3	4	5	6	7	8	9	10
C	A	L	M	E	R	F	O	T	U

3.G. Higher order differentiation

Theory and Examples

Until now, we calculated the first derivative of functions. Here, we will calculate second, and third derivative of a function. These higher derivatives are called higher order derivatives.

We have a function f(x) for which we need to calculate the third derivative. For this we calculate the first derivative$f'(x)$. Then we calculate the derivative of the first derivative $f''(x)$. Then we calculate the derivative of the second derivative $f'''(x)$.

EXAMPLE

Find the third derivative of $f(x) = 2x^3 - x^2 + 3x - 73$

If $f(x) = 2x^3 - x^2 + 3x - 73$
First, we calculate the first derivative of f(x).
$f'(x) = (2x^3)' - (x^2)' + (3x)' - (73)' = 6x^2 - 2x + 3$
Second, we calculate the first derivative of $f'(x)$.
$f''(x) = (6x^2)' - (2x)' + (3)' = 12x - 2$
Third, we calculate the first derivative of $f''(x)$.
$f'''(x) = (12x)' - 2' = 12$

EXAMPLE

Find the third derivative of $f(x) = 3x^3 - e^x$

If $f(x) = 3x^3 - e^x$
First, we calculate the first derivative of f(x)
$f'(x) = (3x^3)' - (e^x)' = 9x^2 - e^x$
Second, we calculate the first derivative of $f'(x)$.
$f''(x) = (9x^2)' - (e^x)' = 18x - e^x$
Third, we calculate the first derivative of $f''(x)$.
$f'''(x) = (18x)' - e^x = 18 - e^x$

Chapter 3. G. Higher order differentiation

Each year, during the festival of Liberalia , in ancient Rome, the freeborn citizens boys from 14 to 17 years old went through the passage from childhood to manhood. What was the month this event took place?

Determine which answer is correct. In the table at the bottom of the page cross off all the letters of the correct answers.

1) If $f(x) = x^3 - 2x^2 + 3x - 4\,;\ then\ f''(x) = 6x - 4$

2) If $f(x) = 7x^3 + 6x^2 + 5x - 4\,;\ then\ f'''(x) = 21x - 6$

3) If $f(x) = 3x^5 + 4\sqrt{x} - 5x - 6\,;\ then\ f'''(x) = 180x^2 + \frac{3}{x^2\sqrt{x}}$

4) If $f(x) = 3\sqrt{5x-1} - 4x + 5\,;\ then\ f''(x) = 18x^2 + \frac{5}{x^3\sqrt{x}}$

5) If $f(x) = \sqrt{7x^2+6x}\,;\ then\ f''(x) = \frac{-36}{(7x^2+6x)\sqrt{7x^2+6x}}$

6) If $f(x) = \frac{2}{2x-3}\,;\ then\ f'''(x) = 2(2x-3)$

7) If $(x) = e^x \ln(x)\,;\ then\ f'''(x) = e^x \ln(x) + 3e^x x^{-1} - 3e^x x^{-2} + 2e^x x^{-3}$

8) If $f(x) = \frac{1}{2x} + \sqrt{x}\,;\ then\ f'''(x) = 2(2x+\sqrt{x})^{-2}$

9) If $f(x) = \sin(x) + x^3\,;\ then\ f''(x) = \sin(x) + 3x$

10) If $f(x) = \cos(x) + e^x\,;\ then\ f''(x) = -\cos(x) + e^x$

1	2	3	4	5	6	7	8	9	10
E	M	B	A	H	R	O	C	H	S

3.H. Implicit differentiation

Theory and Examples

Sometimes we have expressions like $x^2 - y + 3x + y^2 = 5$. In this case we have to differentiate term by term to find the differential $f'(x) = y'$.

EXAMPLE

We have the expression given above. We want to find $f'(x) = y'$.

$x^2 - y + 3x + y^2 = 5$

We differentiate each term of the relation with regard with x. Y is a function of a function.

$(x^2)' - y' + (3x)' + (y^2)' = 5'$

$2x - y' + 3 + 2yy' = 0$

$y'(-1 + 2y) + 2x + 3 = 0$

$y'(-1 + 2y) = -2x - 3$

$y' = \frac{-2x-3}{-1+2y}$

EXAMPLE

If $3x^3 - 3y^3 = 7\ ;\ find\ y''$

We differentiate each term.

$(3x^3)' - (3y^3)' = 7'$

$9x^2 - 9y^2y' = 0$ From here we have: $y' = \frac{x^2}{y^2}$

We differentiate another time each term.

$(9x^2)' - (9y^2y')' = 0$

We apply the product rule for $(9y^2y')'$.

$18x - [18y(y')^2 + 9y^2y''] = 0$

We substitute y' in the equation above.

$18x - 18y(\frac{x^2}{y^2})^2 - 9y^2y'' = 0$

$18x - 18\frac{x^4}{y^3} = 9y^2y''$

$y'' = \frac{2x}{y^2} - \frac{2x^4}{y^5}$

Chapter 3. H. Implicit differentiation

This glass pyramid was completed in 1989. It became a landmark of Paris "the city of lights". The pyramid is situated in the courtyard of the former royal palace now a famous museum called...

Determine which answer is correct. In the table at the bottom of the page cross off all the letters of the correct answers.

1) If $xy + x = 2$; $y' = \frac{-1-y}{x}$

2) If $x^2 + y^2 = 10$; $y' = 2x + 2y$

3) If $2x^2 + y^3 = 3$; $y'' = 4x - 3y^2$

4) If $4x^2 + 2y^2 = 9$; $y'' = 8x + 4y$

5) At point (2,3) the tangent slope to the curve $2x^2 + xy = 2\ is - 5.5$

6) At point (3,4) of the curve $2x^2 + 3xy - y^2 = 38$ the slope of the tangent line is: 2

7) The slope of the tangent line to the graph of $y = \frac{4}{\pi}x - \sin(xy)\ at\ \left(\frac{\pi}{2}, 1\right)\ is: \frac{4}{\pi}$

8) If $3y^2 + \ln(x) = 2y - \cos(x)$, $y' = 6y + \ln(x) + \sin(x)$

9) If $\cot(y) + 2x = 5y - y^2$, $y' = 2x[-csc^2(y) - 5]$

10) If $(y-1)^2 = 6y + x^3 + 2x$, $y' = \frac{3x^2+2}{2(y-1)-6}$

1	2	3	4	5	6	7	8	9	10
B	L	O	U	S	V	U	R	E	C

3.I. Differentiation - Applications

a. Relating graph of f(x) to f ′(x) and f ″(x)

Theory and Examples

We know that the value of the derivative gives the value of the slope of the tangent to the graph. When the second derivative is positive, this means that the slopes of the graph keep increasing. When the second derivative is negative, this means that the slopes of the graph keep decreasing. There are a few rules that we will apply here regarding the relation between the sign of the first and second derivative and the graph of f(x).

	Positive	Negative
First derivative	Graph of f(x) goes UP	Graph of f(x) goes DOWN
Second derivative	Graph is Concave UP	Graph is Concave DOWN

EXAMPLE

Graph the function $f(x) = 2x^3 - 2x^2 - 5x + 4$

First, we calculate the first derivative.

$f'(x) = 6x^2 - 4x - 5$

$f'(x) = 0$ for:

$$x_{1,2} = \frac{4 \mp \sqrt{4^2 - 4(6)(-5)}}{2*6} = \frac{4 \mp \sqrt{16+120}}{12} = \frac{4 \mp \sqrt{136}}{12} = \frac{4 \mp 11.66}{12} \; so, x_1 = -0.63 \; and \; x_2 = 1.3$$

Second, we calculate the second derivative.

$f''(x) = 12x - 4 \; so, f''(x) = 0 \; for \; x = 0.33$

The table below shows the sign of the derivative and the graph behavior.

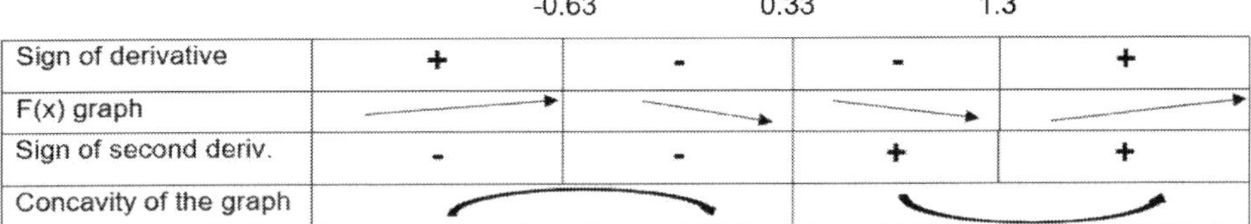

	(to −0.63)	−0.63 to 0.33	0.33 to 1.3	(from 1.3)
Sign of derivative	+	-	-	+
F(x) graph	↗	↘	↘	↗
Sign of second deriv.	-	-	+	+
Concavity of the graph	⌒		‿	

The graph is shown below.

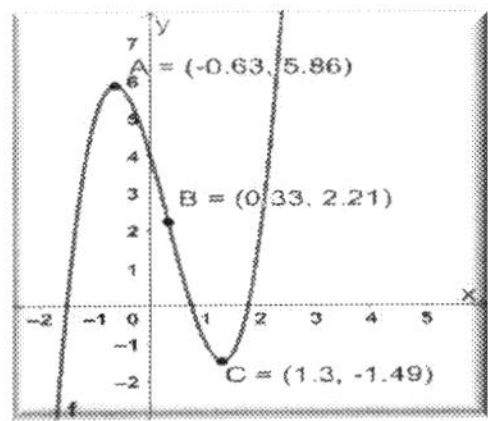

Chapter 3 I. a. Relating graph of f(x) to f '(x) and f ''(x)

This mountain is situated at the Pakistan-China border. It is the second highest mountain in the world after Mount Everest. Its name is…..

Determine which answer is correct. In the table at the bottom of the page cross off all the letters of the correct answers.

1) If the first derivative is positive, the function is increasing.

2) If the second derivative is positive, the function is concave up.

3) The graph of the function $f(x) = \frac{2}{3}x^3 - 2x^2 - 6x + 7$ is increasing between -3 and 1

4) The graph of the function $f(x) = 2x^3 - 7x^2 + 7$ is decreasing between 0 and 2.33

5) The graph of the function $f(x) = \frac{2x-5}{x-1}$ is concave up for values of x less than 1

6) The function $f(x) = \frac{x^2}{x^2-4}$ is concave up between -2 and 1

7) The function $f(x) = \frac{2x}{x+3}$ is concave down between -6 and 6

8) The function $f(x) = \frac{1}{4x^2-9}$ is concave down between x=-1.5 and x=1.5

9) The function $f(x) = \frac{x}{x^2-4}$ is concave up between x=-1 and x=1

10) The function $f(x) = x^5 - 2x^3$ is concave up for $-0.77 \le x \le 0$ *and* $0.77 \le x \le \sqrt{2}$.

1	2	3	4	5	6	7	8	9	10
M	I	K	E	L	T	W	U	O	S

3.I. Differentiation - Applications

b. Differentiability, mean value theorem

Theory and Examples

A function is called differentiable at a point when it has a derivative at that particular point. Remember, the derivative of a function at a point is the instantaneous rate of change at that point. Suppose a continuous function has the graph between f(b) and f(c) where c>b.

The mean value theorem says that there is at least one point on the graph D [d,f(d)] of a continuous and differentiable function, where the slope at that point of the graph equals the slope of the line that passes through the ends points [b,f(b)] and [c,f(c)]. The formula is:

$$f'(d) = \frac{f(c)-f(b)}{c-b}$$

EXAMPLE

We have the graph of the function: $f(x) = (x-1)^5 + 4x^2 - 6x + 3$ showed below for 0<x<1.

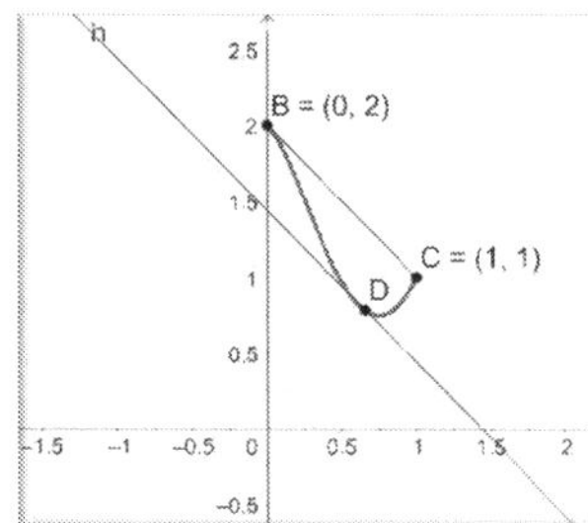

Here, we have the segment that passes through B and C. let's call B point 1 and C point 2.
The slope of the line is calculated:

$$slope_{BC} = \frac{y_2-y_1}{x_2-x_1} = \frac{1-2}{1-0} = -1$$

By the **mean value theorem**, there is a point D on the graph of f(x) where the tangent to the graph has the same slope as the slope of the segment that passes through B and C.

We have to determine the expression of the tangent to the graph. We know that the expression of the slope of the tangent is given by the first derivative of the function.
The derivative of the function is:

$$f'(x) = [(x-1)^5]' + (4x^2)' - (6x)' + 3' = 5(x-1)^4 + 8x - 6$$

We have to find the value of x for which $f'(x) = -1$
In this case this value is x=0.65.

$$f(0.65) = (0.65-1)^5 + 4(0.65)^2 - 6(0.65) + 3 = -0.005 + 1.69 - 3.9 + 3 = 0.78$$

So, the coordinates of point D are: (0.61,0.78)

Chapter 3. I. b. Differentiability, mean value theorem

This mammal once lived all over the Great Plains and much of North America. This animal was very important to the Indigenous peoples that lived in these areas. What animal is this?

Determine which answer is correct. In the table at the bottom of the page cross off all the letters for the correct answers.

1) A differentiable function is a function whose derivative exists all over the domain.

2) In a two-dimensional curve between two points, there is at least one point at which the tangent to the curve is parallel to the secant through its two points.

3) If $f(x) = 3x^2 - 4x + 1$ and the points (-3,40) and, (1,0) the value of x where the tangent at the graph is parallel with the line that goes through (-3,40) and, (1,0) is: 2

4) If $f(x) = x^2 + 5x - 3$ and the points (-1,-7) and, (2,11) the value of x where the tangent at the graph is parallel with the line that goes through (-1,-7) and, (2,11) is: x=4

5) If $f(x) = x^5 - 2x^3$and the points (-1.6,-2.28) and, (1.6,2.28) the values of x where the tangents at the graph are parallel with the line that goes through (-1.6,-2.28) and, (1.6,2.28) are: x=∓1.18

6) If $f(x) = 4x^3 + 3x^2$ the values of x where the tangents of 1 at the graph are: x=-1 and x=5

7) If $f(x) = 6x^4 + 7$ the values of x where the tangents of -3 at the graph are: x=7 and x=9

8) If $f(x) = \frac{1}{x^2} - \frac{2}{5}$ the value of x where the tangent of 2 at the graph is: x=-1

9) If $f(x) = \ln(x) + 10$ the value of x where the tangent of 5 at the graph is: x= $\frac{1}{5}$

10) If $f(x) = x^3 + 2$ and the points (-0.5,1.875) and, (0.5,2.125) the values of x where the tangents at the graph are parallel with the line that goes through (-0.5,1.875) and, (0.5,2.125) are: x=∓4

1	2	3	4	5	6	7	8	9	10
M	E	B	I	U	S	O	C	A	N

3.I. Differentiation - Applications

c. Newton's method

Theory and Examples

In a nutshell, Newton's method approximates the solution of the equation f(x)=0. Let us analyze the figure below.

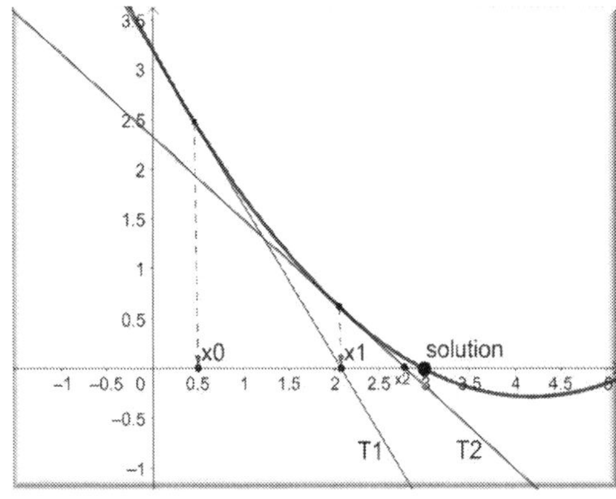

We start by guessing a solution x_0. Here we calculate the tangent at the function.

$f'(x_0) = \frac{f(x)-f(x_0)}{x-x_0}$

This tangent (T_1) intersects the x axis in x_1. x_1 is of coordinates (x_1,0). This point is part of the tangent.

To find x_1, we substitute its coordinates in the tangent relation.

$f'(x_0) = \frac{f(x_1)-f(x_0)}{x_1-x_0}, where\ f(x_1) = 0$

$f'(x_0) * (x_1 - x_0) = -f(x_0)$

$x_1 - x_0 = -\frac{f(x_0)}{f'(x_0)}$

$x_1 = x_0 - \frac{f(x_0)}{f'(x_0)}$

So, $x_2 = x_1 - \frac{f(x_1)}{f'(x_1)}$ and so on. We repeat this process until the value of x_n converges to a value. This value is the solution we are looking for.

EXAMPLE

Find the solution of $f(x) = x^3 + 1$

The values used in the next calculations are shown in the table below

x_i	$f(x)$	$f'(x)$	x_{i+1}
-1.5	-2.37	6.75	-1.1418
-1.1418	-0.51	3.95	-1.018
-1.018	-0.055	3.11	-1.00033

Suppose the solution is -1.5

$f'(x) = 3x^2\ so, f'(-1.5) = 3(-1.5)^2 = 6.75$

$f(-1.5) = -2.37$

$x_1 = x_0 - \frac{f(x_0)}{f'(x_0)} = -1.5 - \frac{-2.37}{6.75} = -1.5 + 0.35 = -1.1418$

$x_2 = x_1 - \frac{f(x_1)}{f'(x_1)} = -1.14 - \frac{-0.51}{3.95} = -1.14 + 0.1298 = -1.018$

$x_3 = x_2 - \frac{f(x_2)}{f'(x_2)} = -1.018 - \frac{-0.055}{3.11} = -1.018 + 0.01797 = -1.00033$

$x_4 = x_3 - \frac{f(x_3)}{f'(x_3)} = -1. - \frac{-0.00098}{3.00196} = -1 + 0.00033 = -1.$

The solution is x=-1

Chapter 3. I. c. Newton's method

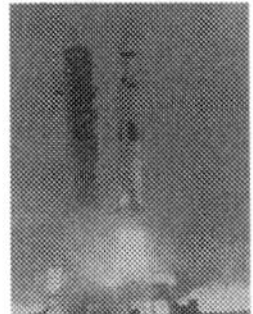

This rocket was built in 1957 using solid fuel upper stages by the Americans to place the satellite Explorer I into high orbit. The rockets name was……

Determine which answer is correct. In the table at the bottom of the page cross off all the letters of the correct answers.

1) The square root of 135, using Newton's method is: 11.61

2) The square root of 432, using Newton's method is: 25.43

3) The solution of $x^3 - 2x^2 = 4$ is: 5

4) The solution of $3x^4 - 4x^3 + 2x = 7$ is: x=1.624

5) The solution of $2x^3 + 3x - 4 = \sin(x)$ is: x=0.5

6) The solution of $x^5 + 4x^3 - 5x = 3$ is: x=1.18

7) The solution of $(x-3)^3 = \sin(x)$ is: x=3.13

8) The solution of $(x-6)^3 = \ln(x)$ is: x=10

9) The solution of $2(x-9)^3 = \ln(x) + 2x$ is: x=11.33

1	2	3	4	5	6	7	8	9
A	J	U	I	N	B	R	O	K

3.I. Differentiation - Applications

d. Problems in contextual situations, including related rates and optimization problems

Theory and Examples

We use math all the time whether we realize it or not. Math helps us build things, we use it in science, at the grocery store, when saving money, for time, etc.

EXAMPLE

At what time does a space shuttle change the trajectory from concave down to concave upward? The equation of the space shuttle height with time for the first 5 minutes is given by:

$h(t) = 2008 - 0.047t^3 + 18.3t^2 - 345t$

Source of the relation: https://spacemath.gsfc.nasa.gov/weekly/5Page40.pdf

A modified equation with different coefficients is showed in the graph below to be able to visualize the trajectory.

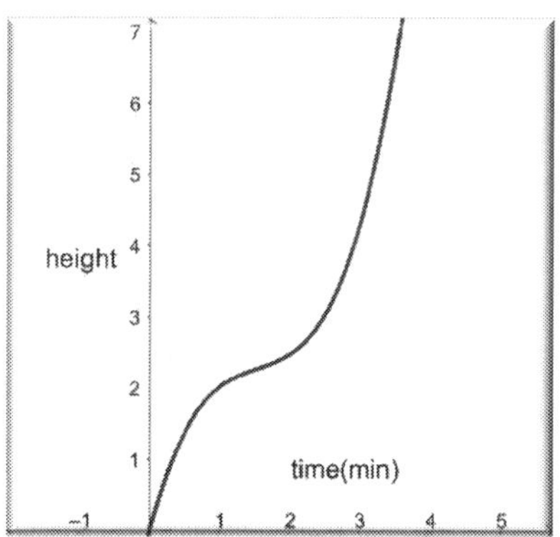

In the calculations we use the real equation used in reality.

To find it we first calculate the first derivative of the equation.

$h'(t) = -3(0.047)t^2 + 2(18.3)t - 345$

Then we calculate the second derivative.

$h''(t) = -3(2)(0.047)t + 2(18.3)$

We equalize the second derivative with zero.

$0 = -3(2)(0.047)t + 2(18.3)$

$0.282t = 36.6$

$t = \frac{36.6}{0.282} = 129.78\, sec = 2.18\, min$

At 2.18 minutes, the shuttle is changing the trajectory from concave down to concave upward.

Chapter 3. I. d. Problems in contextual situations, including related rates and optimization problems

Gaius Marius married the daughter of Gaius Caesar, the grandfather of Julius Caesar. The name of the girl was considered of divine origins in roman times. The girl was called……

Determine which answer is correct. In the table at the bottom of the page cross off all the letters for the correct answers.

For first 4 questions we assume that all variables depend on z.

1) If $x^3 - 2x + y = 1$ the first derivative with respect to z is: $3x^2x' - 2x' + y' = 0$

2) If $2x^5 + 3x^4 - 4x + y^2 = 10$ the first derivative with respect to z is: $10x^4 + 3x^2 - 4 + y^2 = 0$

3) If $x^3 + 3y^2 - x^2 = \sin(x)$ the first derivative with respect to z is: $3x^2x' + 6yy' - 2xx' = x'\cos(x)$

4) If A=πR^2 the first derivative with respect to z is: $A = 2\pi R'$

5) We have a rectangular yard with perimeter of 300 m. It has to be fenced. The dimensions of the yard that will give us the greatest area are: x=y=75m

6) We have to build a box with the length of the base four times the width. The height is the 3 times length minus 10 cm. The dimensions that will minimize the volume are: Length= 2 cm , Width= 3cm , height= 7cm

7) We have to build a box with a base that have length two times the width, and we have 25 m square of material. The dimensions for the maximum volume are: Length= 2.88 m , Width= 1.44 m and Height= 1.93 m

8) We have to build a tunnel that has a cylinder shape that has 50 Liters in volume. The dimensions of the tunnel in order to have the smallest surface area are: R= 4 cm And h = 10 cm

9) The profit relation for a company is $P = Revenue - expenses = \frac{1230}{p} - \frac{550}{(p)^2}$ The price per unit (p) for maximum profit P is: $p = \$0.89$

10) We need to fence two adjacent rectangular lots of land. We have 240 m of fence. The dimensions x and y to maximize the area are: x = 2 m and y = 3 m

1	2	3	4	5	6	7	8	9	10
O	J	E	U	C	L	U	I	M	A

QUICK ANSWERS

Chapter 3

3.B CHINA

3.C POGO

3.D.a WHITE

3.D.b MARIUS

3.E.a STAGING

3.F.a ANDES

3.F.b COWS

3.F.c PERILUNE

3.F.d LEFT

3.G MARCH

3.H LOUVRE

3.I.a K TWO

3.I.b BISON

3.I.c JUNO

3.I.d JULIA

FULL SOLUTIONS

CHAPTER 3

Chapter 3. B. Differentiation-Definition of derivatives

It is known that the formula of the derivative is: $f'(x) = \lim_{h\to 0}\frac{f(x+h)-f(x)}{h}$

Using this formula calculate the following derivatives.

1. Correct

$f(x) = 2x + 45\ so\ f'(x) = 2$

$f'(x) = \lim_{h\to 0}\frac{f(x+h)-f(x)}{h} = \lim_{h\to 0}\frac{2(x+h)+45-2x-45}{h} = \lim_{h\to 0}\frac{2x+2h+45-2x-45}{h} = \lim_{h\to 0}\frac{2h}{h} = \lim_{h\to 0}\frac{2}{1} = 2$

2. Incorrect

$f(x) = 3x^2 + 5x - 4\ so\ f'(x) = 6x - 5$

$f'(x) = \lim_{h\to 0}\frac{f(x+h)-f(x)}{h} = \lim_{h\to 0}\frac{3(x+h)^2+5(x+h)-4-(3x^2+5x-4)}{h} = \lim_{h\to 0}\frac{3(x^2+2hx+h^2)+5x+5h-4-3x^2-5x+4}{h} =$

$\lim_{h\to 0}\frac{3x^2+6hx+3h^2+5x+5h-4-3x^2-5x+4}{h} = \lim_{h\to 0}\frac{6hx+3h^2+5h}{h} = \lim_{h\to 0}\frac{h(6x+3h+5)}{h} = \lim_{h\to 0}(6x + 3h + +5) = 6x + 5$

3. Correct

$f(x) = Ax^2 + Bx - 37\ so\ f'(x) = 2Ax + B$

$f'(x) = \lim_{h\to 0}\frac{f(x+h)-f(x)}{h} = \lim_{h\to 0}\frac{A(x+h)^2+B(x+h)-37-[Ax^2+Bx-37]}{h} = \lim_{h\to 0}\frac{A(x^2+2xh+h^2)+Bx+Bh-37-Ax^2-Bx+37}{h} =$

$\lim_{h\to 0}\frac{Ax^2+2Axh+Ah^2+Bx+Bh-37-Ax^2-Bx+37}{h} = \lim_{h\to 0}\frac{2Axh+Ah^2+Bh}{h} = \lim_{h\to 0}\frac{h(2Ax+Ah+B)}{h} = \lim_{h\to 0}(2Ax + Ah + B) = 2Ax + B$

Using Δx notation instead of h find the derivative of the following functions:

4. Incorrect

$f(x) = 5x^2 - 4x + 3 \; so \; f'(x) = 10x - 4$

$f'(x) = \lim\limits_{\Delta x \to 0} \frac{f(x+\Delta x)-f(x)}{\Delta x} = \lim\limits_{\Delta x \to 0} \frac{5(x+\Delta x)^2-4(x+\Delta x)+3-(5x^2-4x+3)}{\Delta x} =$

$\lim\limits_{\Delta x \to 0} \frac{5[x^2+2x\Delta x+(\Delta x)^2]-4(x+\Delta x)+3-(5x^2-4x+3)}{\Delta x} == \lim\limits_{\Delta x \to 0} \frac{5x^2+10x\Delta x+5(\Delta x)^2-4x-4\Delta x+3-5x^2+4x-3)}{\Delta x} =$

$\lim\limits_{\Delta x \to 0} \frac{10x\Delta x+5(\Delta x)^2-4\Delta x}{\Delta x} = \lim\limits_{\Delta x \to 0} \frac{\Delta x(10x+5\Delta x-4)}{\Delta x} = \lim\limits_{\Delta x \to 0} (10x + 5\Delta x - 4) = 10x - 4$

5. Correct

$f(x) = 7x - 27 \; so \; f'(x) = 7$

$f'(x) = \lim\limits_{\Delta x \to 0} \frac{f(x+\Delta x)-f(x)}{\Delta x} = \lim\limits_{\Delta x \to 0} \frac{7(x+\Delta x)-27-(7x-27)}{\Delta x} = \lim\limits_{\Delta x \to 0} \frac{7x+7\Delta x-27-7x+27}{\Delta x} = \lim\limits_{\Delta x \to 0} \frac{7\Delta x}{\Delta x} = \lim\limits_{\Delta x \to 0} 7 = 7$

6. Incorrect

$f(x) = x^{-1} \; so \; f'(x) = -x^{-2}$

$f'(x) = \lim\limits_{\Delta x \to 0} \frac{f(x+\Delta x)-f(x)}{\Delta x} == \lim\limits_{\Delta x \to 0} \frac{(x+\Delta x)^{-1}-x^{-1}}{\Delta x} == \lim\limits_{\Delta x \to 0} \frac{\frac{1}{x+\Delta x}-x^{-1}}{\Delta x} = \lim\limits_{\Delta x \to 0} \frac{\frac{1}{x+\Delta x}-\frac{x^{-1}(x+\Delta x)}{x+\Delta x}}{\Delta x} =$

$\lim\limits_{\Delta x \to 0} \frac{\frac{1-x^{-1}(x+\Delta x)}{x+\Delta x}}{\Delta x} = \lim\limits_{\Delta x \to 0} \frac{\frac{1-1-x^{-1}\Delta x}{x+\Delta x}}{\Delta x} = \lim\limits_{\Delta x \to 0} \frac{\frac{-x^{-1}\Delta x}{x+\Delta x}}{\Delta x} = \lim\limits_{\Delta x \to 0} \frac{-x^{-1}\Delta x}{\Delta x(x+\Delta x)} = \lim\limits_{\Delta x \to 0} \frac{-x^{-1}}{(x+\Delta x)} = \lim\limits_{\Delta x \to 0} \frac{-x^{-1}}{(x+\Delta x)} = -\frac{x^{-1}}{x} = -x^{-2}$

7. Incorrect

$f(x) = 3x^{-1} + x \; so \; f'(x) = 1 - 3x^{-2}$

$f'(x) = \lim\limits_{\Delta x \to 0} \frac{f(x+\Delta x)-f(x)}{\Delta x} = \lim\limits_{\Delta x \to 0} \frac{3(x+\Delta x)^{-1}+(x+\Delta x)-3x^{-1}-x}{\Delta x} =$

$= \lim\limits_{\Delta x \to 0} \frac{3(x+\Delta x)^{-1}+x+\Delta x-3x^{-1}-x}{\Delta x} = \lim\limits_{\Delta x \to 0} \frac{3(x+\Delta x)^{-1}+\Delta x-3x^{-1}}{\Delta x} = \lim\limits_{\Delta x \to 0} \frac{\frac{3}{x+\Delta x}+\Delta x-3x^{-1}}{\Delta x} = \lim\limits_{\Delta x \to 0} \frac{\frac{3}{x+\Delta x}+\frac{(\Delta x-3x^{-1})(x+\Delta x)}{x+\Delta x}}{\Delta x} =$

$\lim\limits_{\Delta x \to 0} \frac{\frac{3}{x+\Delta x}+\frac{x\Delta x+\Delta x^2-3-3x^{-1}\Delta x}{x+\Delta x}}{\Delta x} = \lim\limits_{\Delta x \to 0} \frac{\frac{x\Delta x+\Delta x^2-3x^{-1}\Delta x}{x+\Delta x}}{\Delta x} = \lim\limits_{\Delta x \to 0} \frac{x\Delta x+\Delta x^2-3x^{-1}\Delta x}{\Delta x(x+\Delta x)} = \lim\limits_{\Delta x \to 0} \frac{\Delta x(x+\Delta x-3x^{-1})}{\Delta x(x+\Delta x)} =$

$\lim\limits_{\Delta x \to 0} \frac{x+\Delta x-3x^{-1}}{(x+\Delta x)} = \frac{x-3x^{-1}}{x} = (x - 3x^{-1})x^{-1} = 1 - 3x^{-2}$

Using the formula $f'(x) = \lim\limits_{h \to 0} \frac{f(x+h)-f(x)}{h}$ find the following derivatives.

8. Correct

$f(x) = x^{-1} + x^2 \; so \; f'(x) = -x^{-2} + 2x = 2x - x^{-2}$

$f'(x) = \lim\limits_{h \to 0} \frac{f(x+h)-f(x)}{h} = \lim\limits_{h \to 0} \frac{(x+h)^{-1}+(x+h)^2-(x^{-1}+x^2)}{h} = \lim\limits_{h \to 0} \frac{\frac{1}{x+h}+(x+h)^2-(x^{-1}+x^2)}{h} =$

$= \lim\limits_{h \to 0} \frac{\frac{1}{x+h}+\frac{(x+h)^3}{x-h}-\frac{(x^{-1}+x^2)(x+h)}{x+h}}{h} = \lim\limits_{h \to 0} \frac{1+(x+h)^3-(1+x^2h+x^3+x^{-1}h)}{h(x+h)} =$

$= \lim_{h\to 0}\frac{1+(x+h)^2(x+h)-(1+x^2h+x^3+x^{-1}h)}{h(x+h)} = \lim_{h\to 0}\frac{1+(x^2+2xh+h^2)\ \ (x+h)-(1+x^2h+x^3+x^{-1}h)}{h(x+h)} =$

$= \lim_{h\to 0}\frac{1+x^3+x^2h+2x^2h+2xh^2+xh^2+h^3-1-x^2h-x^3-x^{-1}h}{h(x+h)} = \lim_{h\to 0}\frac{2x^2h+2xh^2+xh^2+h^3-x^{-1}h}{h(x+h)} =$

$= \lim_{h\to 0}\frac{h(2x^2+3xh+h^2-x^{-1})}{h(x+h)} = \lim_{h\to 0}\frac{2x^2+3xh+h^2-x^{-1}}{x+h} = \frac{2x^2-x^{-1}}{x} = 2x - x^{-2}$

9. Incorrect

$f(x) = x^2 + 3x\ so\ for\ x = 2\ , f'(2) = 7$

$f'(x) = \lim_{h\to 0}\frac{f(x+h)-f(x)}{h} = \lim_{h\to 0}\frac{(x+h)^2+3(x+h)-(x^2+3x)}{h} = \lim_{h\to 0}\frac{x^2+2xh+h^2+3x+3h-x^2-3x}{h} == \lim_{h\to 0}\frac{2xh+h^2+3h}{h} =$

$\lim_{h\to 0}\frac{h(2x+h+3)}{h} = \lim_{h\to 0}(2x+h+3) = 2x+3$

For x=2; $f'(2) = 2(2) + 3 = 7$

10. Correct

$f(x) = x^2 + 33\ so, for\ x = 3\ , f'(3) = 6$

$f'(x) = \lim_{h\to 0}\frac{f(x+h)-f(x)}{h} = \lim_{h\to 0}\frac{(x+h)^2+33-(x^2+33)}{h} = \lim_{h\to 0}\frac{x^2+2xh+h^2+33-x^2-33}{h} = \lim_{h\to 0}\frac{2xh+h^2}{h} =$

$\lim_{h\to 0}\frac{h(2x+h)}{h} = \lim_{h\to 0}(2x+h) = 2x$

For x=3; $f'(3) = 2(3) = 6$

Chapter 3. C. Differentiation-Notation

1. Correct

Gottfried Leibniz's notation is: $\frac{dy}{dx}$

2. Incorrect

Another notation is: $f(x) = y$

3. Correct

Joseph Louis Lagrange's notation is $f'(x)$

4. Incorrect

This is formula for calculating distance d=V*t

5. Correct

Leonhard Euler's notation for the second derivative is: D_x^2

6. Incorrect

There is no such formula F=k/e

7. Correct

One of Isaac Newton's notation is: $\dot{y}$ for first derivative.

8. Incorrect

Henry's formula is: SO=CCER. There is no such formula.

9. Correct

Another formula used by Isaac Newton was: $\dot{x}$

10. Correct

Isaac Newton also used this notation for first derivative: □$\acute{y}$

Chapter 3 D, a. Rate of change- Average versus Instantaneous

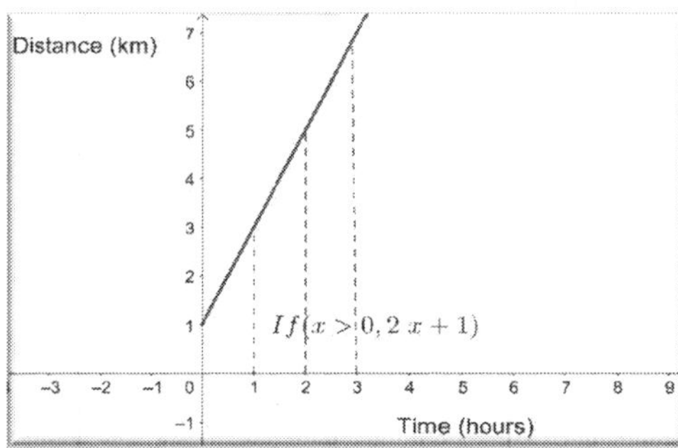

1. Correct

The average rate of change (speed) between 1 and 3 hours is: 2km/h

$speed = \frac{7km-3km}{3hours-1hour} = \frac{4km}{2hours} = 2km/h$

2. Incorrect

The average rate of change (speed) between 0 and 3 hours is: 3km/h

$speed = \frac{7km-1km}{3hours-0hour} = \frac{6km}{3hours} = 2km/h$

A tire rolls by the relation between distance and time as follows: $s(t) = 2t^2 + 4t + 3$

3. Correct

The instantaneous velocity (first derivate) at t=2 seconds is: 12m/s

$s'(t) = 4t + 4, for\ t = 2, s'(2) = 4(2) + 4 = 12m/s$

4. Incorrect

The instantaneous velocity (first derivate) at t=5 seconds is: 24m/s

$s'(t) = 4t + 4, for\ t = 2, s'(5) = 4(5) + 4 = 24m/s$

If $R(t) = \frac{3000t^2+500t}{5} + 20{,}000$ represent the revenue that a company earns in time.

5. Correct

After 10 days, the revenue is increasing with a speed of $32,100 per day.

$R'(t) = \frac{1}{5}(6000t + 500) + 20,000\ so, R'(10) = \frac{1}{5}[6000(10) + 500] + 20,000 = \frac{1}{5}[60,000 + 500] + 20,000 = \frac{1}{5}[60,500] + 20,000 = 12,100 + 20,000 = \$32,100$ per day.

6. Incorrect

After 30 days, the revenue is increasing with a speed of $56,100

$R'(30) = \frac{1}{5}[6000(30) + 500] + 20,000 = \frac{1}{5}[180,000 + 500] + 20,000 = \frac{1}{5}[180,500] + 20,000 = 36,100 + 20,000 = \$56,100$ per day.

The following questions follow the graph below. The graph represents the function:

$f(x) = -0.2x^2 + 2x + 1$

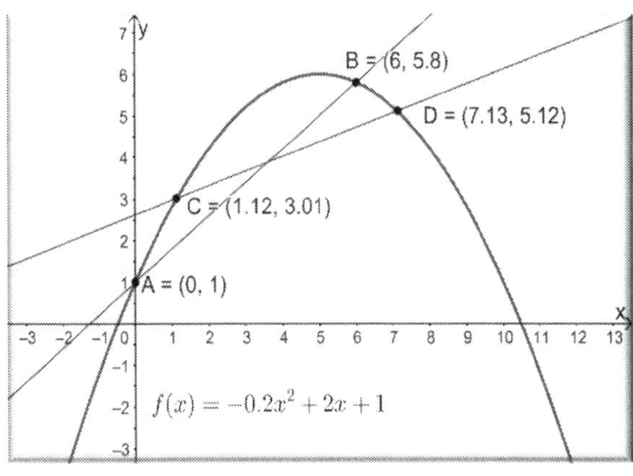

7. Correct

The average rate of change between point A and B is $\frac{5}{6}$

$Av.rate\ of\ change = \frac{5.8-1}{6-0} = \frac{4.8}{6}$

8. Incorrect

The average rate of change between point C and D is not $\frac{6}{5}$

$Av.rate\ of\ change = \frac{5.12-3.01}{7.13-1.12} = \frac{2.11}{6.01}$

9. Correct

The instantaneous rate of change (first derivate) at point C equals 1.552

$f'(x) = -0.2(2)x + 2 = -0.4x + 2\ ;\ f'(1) = -0.4(1.12) + 2 = -0.448 + 2 = 1.552$

10. Incorrect

The instantaneous rate of change (first derivate) at point D equals -0.852

$f'(x) = -0.2(2)x + 2 = -0.4x + 2\ ;\ f'(7) = -0.4(7.13) + 2 = -2.852 + 2 = -0.852$

Chapter 3. D. b. Rate of change- Slope of secant and tangent lines

The next questions follow the graph below. The function represented below is:

$f(x) = x^3 + 3x^2 + x + 2$

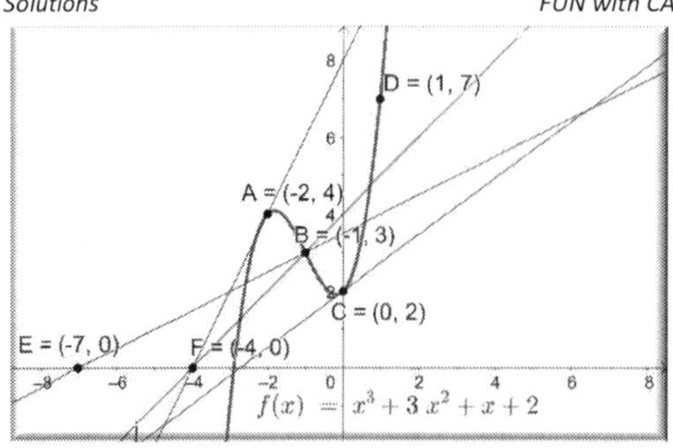

1. Correct

The slope of the secant that passes through the points E and B is: $\frac{1}{2}$

$m = \frac{y_2-y_1}{x_2-x_1} = \frac{3-0}{-1-(-7)} = \frac{3}{6} = \frac{1}{2}$

2. Incorrect

The slope of the secant that passes through the points F and B is 1

$m = \frac{y_2-y_1}{x_2-x_1} = \frac{3-0}{-1-(-4)} = \frac{3}{3} = 1$

3. Correct

The slope of the tangent to the graph in point A equals 1

$f(x) = x^3 + 3x^2 + x + 2\,; so\; f'(x) = 3x^2 + 6x + 1$

$f'(-2) = 3(-2)^2 + 6(-2) + 1 = 12 - 12 + 1 = 1$

4. Incorrect

The slope of the tangent to the graph in point C equals 1

$f(x) = x^3 + 3x^2 + x + 2\,; so\; f'(x) = 3x^2 + 6x + 1$

$f'(0) = 3(0)^2 + 6(0) + 1 = 1$

5. Incorrect

The slope of the tangent to the graph in point D equals 10

$f(x) = x^3 + 3x^2 + x + 2\,; so\; f'(x) = 3x^2 + 6x + 1$

$f'(1) = 3(1)^2 + 6(1) + 1 = 3 + 6 + 1 = 10$

6. Correct

The slope of the tangent to the graph in point B equals -2

$f(x) = x^3 + 3x^2 + x + 2\,; so\; f'(x) = 3x^2 + 6x + 1$

$f'(1) = 3(-1)^2 + 6(-1) + 1 = 3 - 6 + 1 = -2$

7. Incorrect

The slope of the secant that passes through the points A and B is: -1

$m = \frac{y_2-y_1}{x_2-x_1} = \frac{3-4}{-1-(-2)} = \frac{-1}{1} = -1$

8. Correct

The slope of the secant that passes through the points B and C is: -1

$m = \frac{y_2-y_1}{x_2-x_1} = \frac{2-3}{0-(-1)} = \frac{-1}{1} = -1$

9. Incorrect

The slope of the secant that passes through the points B and D is: 2

$m = \frac{y_2 - y_1}{x_2 - x_1} = \frac{7-3}{1-(-1)} = \frac{4}{2} = 2$

10. Incorrect

The slope of the secant that passes through the points C and D is: 5

$m = \frac{y_2 - y_1}{x_2 - x_1} = \frac{7-2}{1-0} = \frac{5}{1} = 5$

Chapter 3. E. Transcendental Functions – Logarithmic, Exponential, Trigonometric

1. Incorrect

If $f(x) = 2e^x - 7\ln(x)\,;\ then\ f'(x) = 2e^x - \frac{7}{x}$

$f'(x) = (2e^x)' - [7\ln(x)]' = 2e^x - \frac{7}{x}$

2. Correct

If $f(x) = 3\ln(x) + 5e^x - ex;\ then\ f'(x) = \frac{3}{x} + 5e^x - e$

$f'(x) = [3\ln(x)]' + [5e^x]' - (ex)' = \frac{3}{x} + 5e^x - e$ (e in the term ex is a constant)

3. incorrect

If $f(x) = 37 - e^x;\ then\ f'(x) = -e^x$

$f'(x) = (37 - e^x)' = -e^x$

4. Incorrect

If $f(x) = 73 + \ln(124) - e^2;\ then\ f'(x) = 0$

$f'(x) = [73 + \ln(124) - e^2]' = 0$ (all the terms are constants)

5. Correct

If $f(x) = -4e^x + \ln(27)\,;\ then\ f'(x) = -4e^x$

$f'(x) = [-4e^x + \ln(27)]' = -(4e^x)' + [\ln(27)]' = -4e^x = -4e^x = -4e^x$

6. Incorrect

If $f(x) = \cos(x) + 3x;\ then\ f'(x) = -\sin(\mathrm{x}) + 3$

$f'(x) = -\sin(x) + 3$

7. Incorrect

If $(x) = 2\ln(x) + 3e^x + \tan(x)\,;\ then\ f'(x) = \frac{2}{x} + 3e^x + sec^2(x)$

$f'(x) = [2\ln(x) + 3e^x + \tan(x)]' = [2\ln(x)]' + [3e^x]' + [\tan(x)]' = \frac{2}{x} + 3e^x + sec^2(x)$

8. Incorrect

If $(x) = \frac{3}{7}\ln(x) + \frac{4}{9}e^x + \frac{5}{9}\tan(x)$; $then\ f'(x) = \frac{3}{7x} + \frac{4}{9}e^x + \frac{5}{9}sec^2(x)$

$f'(x) = [\frac{3}{7}\ln(x) + \frac{4}{9}e^x + \frac{5}{9}\tan(x)]' = [\frac{3}{7}\ln(x)]' + [\frac{4}{9}e^x]' + [\frac{5}{9}\tan(x)]' = \frac{3}{7x} + \frac{4}{9}e^x + \frac{5}{9}sec^2(x)$

9. Correct

If $(x) = \frac{5}{9}\ln(2) + \frac{1}{2}e^x + \cos(x)$; $then\ f'(x) = \frac{1}{2}e^x - \sin(x)$

$f'(x) = [\frac{5}{9}\ln(2) + \frac{1}{2}e^x + \cos(x)]' = [\frac{5}{9}\ln(2)]' + [\frac{1}{2}e^x]' + [\cos(x)]' = 0 + \frac{1}{2}e^x - \sin(x)$

10. Incorrect

If $(x) = 4e^x + \frac{5}{6}\ln(x) - 6\cot(x)$; $then\ f'(x) = 4e^x + \frac{5}{6x} + 6csc^2(x)$

$f'(x) = [4e^x + \frac{5}{6}\ln(x) - 6\cot(x)]' = [4e^x]' + [\frac{5}{6}\ln(x)]' - 6\cot(x)]' = 4e^x + \frac{5}{6x} - 6[-csc^2(x)] = 4e^x + \frac{5}{6x} + 6csc^2(x)$

Chapter 3. F. a. Differentiation rules - Power

1. Correct

If $f(x) = 3x^2 - 7x + 89\ then\ f'(x) = 6x - 7$

$f'(x) = 2(3)x^{2-1} - 7x^{1-1} = 6x - 7x^0 = 6x - 7(1) = 6x - 7$

2. Incorrect

If $f(x) = \frac{2}{5}x^2 + 27x - 63$

$f'(x) = 2\left(\frac{2}{5}\right)x^{2-1} + 27x^{1-1} - 0 = \left(\frac{4}{5}\right)x^2 + 27x^0 = \frac{4}{5}x + 27$

3. incorrect

If $f(x) = \frac{3}{7}x^4 - 2x^3 - 6x^2 + 81$

$f'(x) = 4(\frac{3}{7})x^{4-1} - 3(2)x^{3-1} - 2(6)x^{2-1} + 0 = \frac{12}{7}x^3 - 6x^2 - 12x$

4. Correct

If $f(x) = x^4 - 2x^3 - 6x^2 + 81\ then\ f'(x) = 4x^3 - 6x^2 - 12x$

$f'(x) = 4x^{4-1} - 2(3)x^{3-1} - 2(6)x^{2-1} + 0 = 4x^3 - 6x^2 - 12x$

5. Incorrect

If $f(x) = \frac{1}{x}$

Using the exponents rules we have: $f(x) = \frac{1}{x} = x^{-1}$

$So\ f'(x) = -1x^{-1-1} = -x^{-2} = -\frac{1}{x^2}$

6. Correct

If $f(x) = \frac{1}{x^4}\ then\ f'(x) = -4x^{-5}$

Using the exponents rules we have: $f(x) = \frac{1}{x^4} = x^{-4}$

$So\ f'(x) = -4x^{-4-1} = -4x^{-5} = \frac{-4}{x^5}$

7. Incorrect

If $f(x) = \sqrt{x^3}$

Using the exponents rules we have: $f(x) = \sqrt{x^3} = x^{\frac{3}{2}}$

$So\ f'(x) = \frac{3}{2}x^{\frac{3}{2}-1} = \frac{3}{2}x^{\frac{1}{2}} = \frac{3}{2}\sqrt{x}$

8. Incorrect

If $f(x) = \frac{\sqrt{x}}{x^4}$

Using the exponents rules we have: $f(x) = \frac{\sqrt{x}}{x^4} = x^{\frac{1}{2}} * x^{-4} = x^{\frac{1}{2}-4} = x^{\frac{1}{2}-4} = x^{-\frac{7}{2}}$

$So\ f'(x) = -\frac{7}{2}x^{-\frac{7}{2}-1} = -\frac{7}{2}x^{-\frac{9}{2}} = -\frac{7}{2} * \frac{1}{x^{\frac{9}{2}}} = -\frac{7}{2\sqrt{x^9}} = -\frac{7}{2x^4\sqrt{x}}$

9. Correct

If $f(x) = \frac{\sqrt{x}}{x^2} + 3x\ then\ f'(x) = -\frac{7}{2x^2\sqrt{x}} + 3$

Using the exponents rules we have: $f(x) = \frac{\sqrt{x}}{x^2} + 3x = x^{\frac{1}{2}} * x^{-2} + 3x = x^{\frac{1}{2}-2} = x^{-\frac{3}{2}} + 3x$

$So\ f'(x) = -\frac{3}{2}x^{-\frac{3}{2}-1} + 3 = -\frac{3}{2}x^{-\frac{5}{2}} + 3 = -\frac{3}{2} * \frac{1}{x^{\frac{5}{2}}} + 3 = -\frac{3}{2\sqrt{x^5}} + 3 = -\frac{3}{2x^2\sqrt{x}} + 3$

10. Correct

If $f(x) = \frac{x^3}{x^{-2}}\ then\ f'(x) = 5x^4$

Using the exponents rules we have: $f(x) = \frac{x^3}{x^{-2}} = x^3 * x^2 = x^5$

$So\ f'(x) = 5x^{5-1} = 5x^4$

Chapter 3. F. b. Differentiation rules – Product

1. Correct

If $f(x) = 2x^2\cos(x); then\ f'(x) = 4x\cos(x) - 2x^2\sin(x)$

$f'(x) = (2x^2)' \cos(x) + 2x^2[\cos(x)]' = 4xcos(x) + 2x^2[-\sin(x)] = 4xcos(x) - 2x^2 \sin(x)$

2. Incorrect

If $f(x) = 3x^4\sqrt{x}; then\ f'(x) = 12\frac{3}{2}x^3\sqrt{x}$

$f(x) = 3x^4\sqrt{x}; then\ f'(x) = (3x^4)'\sqrt{x} + 3x^4(\sqrt{x})' = 12x^3\sqrt{x} + 3x^4\frac{1}{2}x^{-\frac{1}{2}} = 12x^3\sqrt{x} + \frac{3}{2}\frac{x^4}{\sqrt{x}} = 12x^3\sqrt{x} + \frac{3}{2}\frac{x^4}{\sqrt{x}}\frac{\sqrt{x}}{\sqrt{x}} = 12x^3\sqrt{x} + \frac{3}{2}x^3\sqrt{x} = 12\frac{3}{2}x^3\sqrt{x}$

$Where;\ (\sqrt{x})' = [(x)^{\frac{1}{2}}]' = \frac{1}{2}x^{-\frac{1}{2}}$

3. Correct

If $f(x) = 4x^2 \tan(x)\,; then\ f'(x) = (8x)\tan(x) + 4x^2sec^2(x)$

$f'(x) = (4x^2)' \tan(x) + 4x^2[\tan(x)]' = (8x)tan(x) + 4x^2sec^2(x)$

4. Incorrect

If $f(x) = \frac{2}{5}x^2e^x; then\ f'(x) = \frac{2}{5}xe^x(2 + x)$

$f'(x) = (\frac{2}{5}x^2)'e^x + \frac{2}{5}x^2[e^x]' = \frac{4}{5}xe^x + \frac{2}{5}x^2e^x = \frac{2}{5}xe^x(2 + x)$

5. Correct

If $f(x) = 2^x \sin(x)\,; then\ f'(x) = 2^x \ln 2 + 2^x \cos(x)$

$f'(x) = (2^x)' \sin(x) + 2^x[\sin(x)]' = 2^x \ln 2 \sin(x) + 2^x \cos(x)$

6. Incorrect

If $f(x) = 3\ln|x|x^2; then\ f'(x) = 3x + 6x\ln|x|$

$f'(x) = (3\ln|x|)'x^2 + 3\ln|x|(x^2)' = \frac{3}{x} * x^2 + 3\ln|x| * 2x = 3x + 6x\ln|x|$

7. Correct

If $f(x) = \ln|x|\sqrt{x}; then\ f'(x) = x^{-\frac{1}{2}}(1 + \frac{1}{2}\ln|x|)$

$f'(x) = (\ln|x|)'\sqrt{x} + \ln|x|(\sqrt{x})' = \frac{\sqrt{x}}{x} + \ln|x|\frac{1}{2}x^{-\frac{1}{2}} = x^{\frac{1}{2}-1} + \ln|x|\frac{1}{2}x^{-\frac{1}{2}} = x^{-\frac{1}{2}}(1 + \frac{1}{2}\ln|x|)$

$Where;\ (\sqrt{x})' = [(x)^{\frac{1}{2}}]' = \frac{1}{2}x^{-\frac{1}{2}}$

8. Incorrect

If $f(x) = \cos(x) * e^x; then\ f'(x) = e^x[\cos(x) - \sin(x)]$

$f'(x) = [\cos(x)]'e^x + \cos(x)[e^x]' = -\sin(x)e^x + \cos(x)e^x = e^x[\cos(x) - \sin(x)]$

9. Correct

If $f(x) = e^x3^x; then\ f'(x) = e^x3^x[1 + \ln(3)]$

$f'(x) = [e^x]'3^x + e^x[3^x]' = e^x3^x + e^x3^x\ln(3) = e^x3^x[1 + \ln(3)]$

10. Correct

If $f(x) = e^x x^5; then\ f'(x) = e^x x^4(x + 5)$

$f'(x) = [e^x]'x^5 + e^x[x^5]' = e^x x^5 + e^x * 5x^4 = e^x x^4(x + 5)$

Chapter 3. F. c. Differentiation rules – Quotient

1. Correct

If $f(x) = \tan(x); then\ f'(x) = [\sec(x)]^2$

$f(x) = \frac{\sin(x)}{\cos(x)}$

$f'(x) = \frac{[\sin(x)]'\cos(x) - \sin(x)[\cos(x)]'}{[\cos(x)]^2} = \frac{\cos(x)\cos(x) - \sin(x)[-\sin(x)]}{[\cos(x)]^2} = \frac{[\cos(x)]^2 + [\sin(x)]^2}{[\cos(x)]^2} = \frac{1}{[\cos(x)]^2} = [\sec(x)]^2$

2. Incorrect

If $f(x) = \frac{\sin(x)}{x^2}; then\ f'(x) = \frac{[xcos(x) - 2\sin(x)]}{x^3}$

$f'(x) = \frac{[\sin(x)]'x^2 - \sin(x)[x^2]'}{[x^2]^2} = \frac{x^2\cos(x) - 2xsin(x)}{x^4} = \frac{x[xcos(x) - \sin(x)2]}{x^4} = \frac{[xcos(x) - 2\sin(x)]}{x^3}$

3. incorrect

If $f(x) = \frac{x}{e^x}; then\ f'(x) = \frac{1-x}{e^x}$

$f'(x) = \frac{x'e^x - x[e^x]'}{[e^x]^2} = \frac{e^x - xe^x}{[e^x]^2} = \frac{e^x(1-x)}{[e^x]^2} = \frac{1-x}{e^x}$

4. Incorrect

If $f(x) = \frac{\sin(x)}{x^2+1}; then\ f'(x) = \frac{\cos(x)(x^2+1) - \sin(x)*2x}{[x^2+1]^2}$

$f'(x) = \frac{[\sin(x)]'(x^2+1) - \sin(x)[x^2+1]'}{[x^2+1]^2} = \frac{\cos(x)(x^2+1) - \sin(x)*2x}{[x^2+1]^2}$

5. Incorrect

If $f(x) = \frac{x^2+2x}{\sqrt{x}}; then\ f'(x) = \frac{\sqrt{x}(1.5x-3)}{x}$

$f'(x) = \frac{[x^2+2x]'\sqrt{x} - (x^2+2x)(\sqrt{x})'}{(\sqrt{x})^2} = \frac{(2x+2)\sqrt{x} - (x^2+2x)\frac{1}{2\sqrt{x}}}{x}$

$(\sqrt{x})' = (x^{\frac{1}{2}})' = \frac{x^{\frac{1}{2}-1}}{2} = \frac{x^{-\frac{1}{2}}}{2} = \frac{1}{2\sqrt{x}}$

$f'(x) = \frac{(2x+2)\sqrt{x} - (x^2+2x)\frac{x^{-\frac{1}{2}}}{2}}{x} = \frac{2x\sqrt{x} - 2\sqrt{x} - \frac{x^{\frac{3}{2}}}{2} - \frac{2x^{\frac{1}{2}}}{2}}{x} = \frac{2x\sqrt{x} - 2\sqrt{x} - \frac{x\sqrt{x}}{2} - \sqrt{x}}{x} = \frac{\sqrt{x}(2x - 2 - 0.5x - 1)}{x} = \frac{\sqrt{x}(1.5x-3)}{x}$

6. Correct

If $f(x) = 2x\sec(x) = \frac{2x}{cox(x)}; then\ f'(x) = \frac{2[1-xtan(x)]}{\cos(x)}$

$f'(x) = \frac{(2x)'\cos(x)-2x[\cos(x)]'}{[\cos(x)]^2} = \frac{2\cos(x)-2x[-\sin(x)]}{[\cos(x)]^2} = \frac{2\cos(x)+2xsin(x)}{[\cos(x)]^2} = \frac{2[\cos(x)-xsin(x)]}{[\cos(x)]^2} = \frac{2\cos(x)[1-\frac{xsin(x)}{\cos(x)}]}{[\cos(x)]^2} = \frac{2[1-xtan(x)]}{\cos(x)}$

7. Incorrect

If $f(x) = \frac{3x^2-4x+2}{\sin(x)}; then\ f'(x) = \frac{(6x-4)\sin(x)-(3x^2-4x+2)\cos(x)}{[\sin(x)]^2}$

$f'(x) = \frac{(3x^2-4x+2)'\sin(x)-(3x^2-4x+2)[\sin(x)]'}{[\sin(x)]^2} = \frac{(6x-4)\sin(x)-(3x^2-4x+2)\cos(x)}{[\sin(x)]^2}$

8. Incorrect

$f(x) = \frac{x^2+2x-3}{5x+4}; then\ f'(x) = \frac{5x^2+8x+23}{(5x+4)^2}$

$f'(x) = \frac{(x^2+2x-3)'(5x+4)-(x^2+2x-3)(5x+4)'}{(5x+4)^2} = \frac{(2x+2)(5x+4)-(x^2+2x-3)*5}{(5x+4)^2} = \frac{10x^2+8x+10x+8-5x^2-10x+15}{(5x+4)^2} = \frac{5x^2+8x+23}{(5x+4)^2}$

9. Incorrect

If $f(x) = \frac{x^2}{lnx}; then\ f'(x) = \frac{x(2lnx-1)}{[\ln(x)]^2}$

$f'(x) = \frac{(x^2)'\ln(x)-x^2[\ln(x)]'}{[\ln(x)]^2} = \frac{(2x)lnx-\frac{x^2}{x}}{[\ln(x)]^2} = \frac{(2x)lnx-x}{[\ln(x)]^2} = \frac{x(2lnx-1)}{[\ln(x)]^2}$

10. Incorrect

$f(x) = \frac{2x^2-4x+6}{2x}; then\ f'(x) = 1 - \frac{3}{x^2}$

$f'(x) = \frac{(2x^2-4x+6)'2x-(2x^2-4x+6)*2}{4x^2} = \frac{(4x-4)*2x-4x^2+8x-12}{4x^2} = \frac{8x^2-8x-4x^2+8x-12}{4x^2} = \frac{4x^2-12}{4x^2} = \frac{4(x^2-3)}{4x^2} = \frac{x^2-3}{x^2} = 1 - \frac{3}{x^2}$

Chapter 3. F. d. Differentiation rules – Chain

1. Correct

If $f(x) = (x^2-4x+5)^2; then\ f'(x) = 2(x^2-4x+5)(2x-4)$

$f'(x) = [x^2-4x+5)^2]'(x^2-4x+5)' = 2(x^2-4x+5)(2x-4)$

2. Correct

If $f(x) = (3x^2+7x)^{\frac{3}{2}}; then\ f'(x) = \frac{3}{2}\sqrt{3x^2+7x}(6x+7)$

$f'(x) = \frac{3}{2}(3x^2 + 7x)^{\frac{3}{2}-1}(3x^2 + 7x)' = \frac{3}{2}(3x^2 + 7x)^{\frac{1}{2}}(6x + 7) = \frac{3}{2}\sqrt{3x^2 + 7x}\,(6x + 7)$

3. incorrect

If $f(x) = \sqrt{x^3 - 5x}; then\ f'(x) = \frac{(3x^2-5)}{2\sqrt{x^3-5x}}$

$f(x) = \sqrt{x^3 - 5x} = (x^3 - 5x)^{\frac{1}{2}}$ so:

$f'(x) = \frac{1}{2}(x^3 - 5x)^{\frac{1}{2}-1}(x^3 - 5x)' = \frac{1}{2}(x^3 - 5x)^{\frac{1}{2}-1}(3x^2 - 5) = \frac{1}{2}(x^3 - 5x)^{-\frac{1}{2}}(3x^2 - 5) = \frac{(3x^2-5)}{2\sqrt{x^3-5x}}$

4. Correct

If $f(x) = \sqrt{e^x - 2x}; then\ f'(x) = \frac{(e^x-2)}{2\sqrt{(e^x-2x}}$

$f(x) = \sqrt{e^x - 2x} = (e^x - 2x)^{\frac{1}{2}}$ so:

$f'(x) = [(e^x - 2x)^{\frac{1}{2}}]'(e^x - 2x)' = \frac{1}{2}(e^x - 2x)^{\frac{1}{2}-1}(e^x - 2) = \frac{1}{2}(e^x - 2x)^{-\frac{1}{2}}(e^x - 2) = \frac{(e^x-2)}{2\sqrt{(e^x-2x}}$

5. Incorrect

If $f(x) = \sqrt{\sin(x)}; then\ f'(x) = \frac{1}{2}\frac{\cos(x)}{2\sqrt{\sin(x)}}$

$f(x) = \sqrt{\sin(x)} = [\sin(x)]^{\frac{1}{2}}$

$f'(x) = \frac{1}{2}[\sin(x)]^{\frac{1}{2}-1}[\sin(x)]' = \frac{1}{2}[\sin(x)]^{-\frac{1}{2}}\cos(x) = \frac{\cos(x)}{2\sqrt{\sin(x)}}$

6. Correct

If $f(x) = 2[\ln(4x - 3)]; then\ f'(x) = \frac{8}{4x-3}$

$f'(x) = 2[\ln(4x - 3)]'(4x - 3)' = \frac{2(4)}{4x-3} = \frac{8}{4x-3}$

7. Incorrect

If $f(x) = [\ln(e^x - e^{-x})]; then\ f'(x) = \frac{e^x+e^{-x}}{(e^x-e^{-x})}$

$f'(x) = [\ln(e^x - e^{-x})]'(e^x - e^{-x})' = \frac{e^x+e^{-x}}{(e^x-e^{-x})}$

8. Correct

If $f(x) = \sin(x - 1)\ ;\ then\ f'(x) = \cos(x - 1)$

$f'(x) = \cos(x - 1)(x - 1)' = \cos(x - 1)$

9. Incorrect

If $f(x) = e^x + 2\ln(x + 7)\ ;\ then\ f'(x) = e^x + \frac{2}{x+7}$

$f'(x) = (e^x)' + [2\ln(x + 7)]'(x + 7)' = e^x + \frac{2}{x+7}$

10. Correct

If $f(x) = \ln(x^2 - 4x)$; $then\ f'(x) = \frac{2(x-2)}{x(x-4)}$

$f'(x) = [\ln(x^2 - 4x)]'(x^2 - 4x)' = \frac{2x-4}{x^2-4x} = \frac{2(x-2)}{x(x-4)}$

Chapter 3. G. Higher order differentiation

1. Correct

If $f(x) = x^3 - 2x^2 + 3x - 4$; $then\ f''(x) = 6x - 4$

$f'(x) = 3x^2 - 4x + 3$

$f''(x) = 6x - 4$

2. Incorrect

If $f(x) = 7x^3 + 6x^2 + 5x - 4$; $then\ f'''(x) = 42$

$f'(x) = 21x^2 + 12x + 5$

$f''(x) = 42x + 12$

$f'''(x) = 42$

3. Correct

If $f(x) = 3x^5 + 4\sqrt{x} - 5x - 6$; $then\ f'''(x) = 180x^2 + \frac{3}{x^2\sqrt{x}}$

$f'(x) = 15x^4 + 4x^{\frac{1}{2}-1} - 5 = 15x^4 + 4x^{-\frac{1}{2}} - 5$

$f''(x) = 60x^3 + 4\left(-\frac{1}{2}\right)x^{-\frac{1}{2}-1} = 60x^3 - 2x^{-\frac{3}{2}}$

$f'''(x) = 180x^2 - 2\left(-\frac{3}{2}\right)x^{-\frac{3}{2}-1} = 180x^2 + (3)x^{-\frac{3}{2}-1} = 180x^2 + (3)x^{-\frac{5}{2}} = 180x^2 + \frac{3}{x^2\sqrt{x}}$

4. Incorrect

If $f(x) = 3\sqrt{5x-1} - 4x + 5$; $then\ f''(x) = \frac{15}{2\sqrt{5x-1}} - \frac{75x^2}{4(5x-1)\sqrt{5x-1}}$

We apply the chain rule for differentiating $\sqrt{5x-1}$, where $u(x) = 5x - 1\ and\ f(u) = \sqrt{u}$

$f'(x) = [3(5x-1)^{\frac{1}{2}}]' - (4x)' = 3[\frac{1}{2}(5x-1)^{\frac{1}{2}-1}](5x-1)' - 4 = \frac{3}{2}(5x-1)^{-\frac{1}{2}}(5x) - 4 = \frac{15x}{2}(5x-1)^{-\frac{1}{2}} - 4$

$f''(x) = [\frac{15x}{2}(5x-1)^{-\frac{1}{2}}]' - (4)' = \left(\frac{15x}{2}\right)'(5x-1)^{-\frac{1}{2}} + \frac{15x}{2}[(5x-1)^{-\frac{1}{2}}]' = \frac{15}{2}(5x-1)^{-\frac{1}{2}} + \frac{15x}{2}[-\frac{1}{2}(5x-1)^{-\frac{1}{2}-1}(5x-1)' = \frac{15}{2}(5x-1)^{-\frac{1}{2}} + \frac{15x}{2}[-\frac{1}{2}(5x-1)^{-\frac{3}{2}}(5x) = \frac{15}{2}(5x-1)^{-\frac{1}{2}} - \frac{75x^2}{4}(5x-1)^{-\frac{3}{2}} = \frac{15}{2\sqrt{5x-1}} - \frac{75x^2}{4(5x-1)\sqrt{5x-1}}$

Where:

For $[\frac{15x}{2}(5x-1)^{-\frac{1}{2}}]'$ we applied the product rule

For $[(5x-1)^{-\frac{1}{2}}]'$ we applied the chain rule

5. Correct

If $f(x)=\sqrt{7x^2+6x}\ ;\ then\ f''(x)=\frac{-36}{(7x^2+6x)\sqrt{7x^2+6x}}$

$f(x)=\sqrt{7x^2+6x}=(7x^2+6x)^{\frac{1}{2}}$

$f'(x)=[(7x^2+6x)^{\frac{1}{2}}]'=\frac{1}{2}(7x^2+6x)^{\frac{1}{2}-1}(7x^2+6x)'=\frac{1}{2}(7x^2+6x)^{-\frac{1}{2}}(14x+6)$

Where

For $[(7x^2+6x)^{\frac{1}{2}}]'$ we applied chain rule.

$f''(x)=[\frac{1}{2}(7x^2+6x)^{-\frac{1}{2}}]'(14x+6)+\frac{1}{2}(7x^2+6x)^{-\frac{1}{2}}(14x+6)'=-\frac{1}{4}(7x^2+6x)^{-\frac{1}{2}-1}(7x^2+6x)'(14x+6)+\frac{1}{2}(7x^2+6x)^{-\frac{1}{2}}(14)=-\frac{1}{4}(7x^2+6x)^{-\frac{3}{2}}(14x+6)(14x+6)+7(7x^2+6x)^{-\frac{1}{2}}=-\frac{(14x+6)^2}{4(7x^2+6x)\sqrt{7x^2+6x}}+\frac{7}{\sqrt{7x^2+6x}}=\frac{28(7x^2+6x)-(14x+6)^2}{(7x^2+6x)\sqrt{7x^2+6x}}=\frac{-36}{(7x^2+6x)\sqrt{7x^2+6x}}$

Where

For $[\frac{1}{2}(7x^2+6x)^{-\frac{1}{2}}]'$ we applied the chain rule.

6. Incorrect

If $f(x)=\frac{2}{2x-3}\ ;\ then\ f'''(x)=\frac{-48}{(2x-3)^4}$

$f(x)=\frac{2}{2x-3}=2(2x-3)^{-1}$

$f'(x)=-2(2x-3)^{-2}(2x-3)'=-2(2x-3)^{-2}(2)=-4(2x-3)^{-2}$

We applied the chain rule for $(2x-3)^{-1}$

$f''(x)=(-4)[(2x-3)^{-2}]'=-4(-2)(2x-3)^{-3}(2x-3)'=8(2x-3)^{-3}(2)=16(2x-3)^{-3}$

$f'''(x)=[16(2x-3)^{-3}]'=16(-3)(2x-3)^{-4}(2x-3)'=-24(2x-3)^{-4}(2)=\frac{-48}{(2x-3)^4}$

We applied again the chain rule.

7. Correct

If $(x)=e^x\ln(x)\ ;\ then\ f'''(x)=e^x\ln(x)+3e^xx^{-1}-3e^xx^{-2}+2e^xx^{-3}$

We are applying the product rule.

$f'(x)=(e^x)'\ln(x)+e^x[\ln(x)]'=e^x\ln(x)+e^xx^{-1}$

$f''(x)=[e^x]'\ln(x)+e^xx^{-1}+[e^x]'x^{-1}+e^x[x^{-1}]'=e^x\ln(x)+e^xx^{-1}+e^xx^{-1}-e^xx^{-2}=e^x\ln(x)+2e^xx^{-1}-e^xx^{-2}$

$f'''(x)=[e^x]'\ln(x)+e^x[\ln(x)]'+2[e^x]'x^{-1}+2e^x[x^{-1}]'-[e^x]'x^{-2}-e^x[x^{-2}]'=e^x\ln(x)+e^xx^{-1}+2e^xx^{-1}-2e^xx^{-2}-e^xx^{-2}-e^x(-2)x^{-3}=e^x\ln(x)+3e^xx^{-1}-3e^xx^{-2}+2e^xx^{-3}$

8. Incorrect

If $f(x) = \frac{1}{2x} + \sqrt{x}$; $then\ f'''(x) = \frac{-3}{x^4} + \frac{3}{8x^2\sqrt{x}}$

$f(x) = \frac{1}{2x} + \sqrt{x} = \frac{1}{2}x^{-1} + x^{\frac{1}{2}}$

$f'(x) = -\frac{1}{2}x^{-2} + \frac{1}{2}x^{\frac{1}{2}-1} = -\frac{1}{2}x^{-2} + \frac{1}{2}x^{-\frac{1}{2}}$

$f''(x) = x^{-3} - \frac{1}{4}x^{-\frac{1}{2}-1} = x^{-3} - \frac{1}{4}x^{-\frac{3}{2}}$

$f'''(x) = -3x^{-4} + \frac{3}{8}x^{-\frac{3}{2}-1} = -3x^{-4} + \frac{3}{8}x^{-\frac{5}{2}} = \frac{-3}{x^4} + \frac{3}{8x^2\sqrt{x}}$

9. Incorrect

If $f(x) = \sin(x) + x^3$; $then\ f''(x) = -\sin(x) + 6x$

$f'(x) = [\sin(x)]' + (x^3)' = \cos(x) + 3x^2$

$f''(x) = [\cos(x)]' + (3x^2)' = -\sin(x) + 6x$

10. Correct

If $f(x) = \cos(x) + e^x$; $then\ f''(x) = -\cos(x) + e^x$

$f'(x) = [\cos(x)]' + (e^x)' = -\sin(x) + e^x$

$f''(x) = [-\sin(x)]' + (e^x)' = -\cos(x) + e^x$

Chapter 3. H. Implicit differentiation

1. Correct

If $xy + x = 2$; $y' = \frac{-1-y}{x}$

Using implicit differentiation, we have:

$(xy)' + x' = 0$

$(x)'y + xy' + 1 = 0$

$y + xy' = -1$

$xy' = -1 - y, so\ y' = \frac{-1-y}{x}$

2. Incorrect

If $x^2 + y^2 = 10$; $y' = -\frac{x}{y}$

Using implicit differentiation, we have:

$(x^2)' + (y^2)' = 0$

$2x + 2y(y)' = 0$

$2y(y)' = -2x, so\ y' = \frac{-2x}{2y} = -\frac{x}{y}$

3. incorrect

If $2x^2 + y^3 = 3$; $y'' = \frac{-12y^3-32x^2}{9y^5}$

Using implicit differentiation, we have:

$(2x^2)' + (y^3)' = 0$

$4x + 3y^2(y)' = 0$

$3y^2(y)' = -4x, so\ y' = \frac{-4x}{3y^2}$

Then, using the quotient rule, we have:

$$y'' = \frac{(-4x)'(3y^2)-(-4x)(3y^2)'}{(3y^2)^2} = \frac{-4(3y^2)+4x(6yy')}{9y^4} = \frac{-12y^2+24xy(\frac{-4x}{3y^2})}{9y^4} = \frac{-12y^2-\frac{32x^2}{y}}{9y^4} = \frac{-12y^3-32x^2}{9y^5}$$

4. Incorrect

If $4x^2 + 2y^2 = 9$; $y'' = \frac{-2y^2-4x^2}{y^3}$

Using implicit differentiation, we have:

$(4x^2)' + (2y^2)' = 0$

$8x + 4y(y)' = 0$

$4y(y)' = -8x, so, y' = \frac{-8x}{4y} = -\frac{2x}{y}$

Then, using the quotient rule, we have:

$$y'' = \frac{(-2x)'y-(-2x)\,y'}{y^2} = \frac{-2y+2x(\frac{-2x}{y})}{y^2} = \frac{-2y^2-4x^2}{y^3}$$

5. Correct

At point (2,3) the tangent slope to the curve $2x^2 + xy = 2\ is - 5.5$

Using implicit differentiation, we have:

$(2x^2)' + (xy)' = 0$

$4x + (x)'y + x(y)' = 0$

$4x + y + x(y)' = 0$

$x(y)' = -4x - y\ so, y' = \frac{-4x-y}{x}$

We substitute x=2 and y=3, so we have:

$y' = \frac{-4(2)-3}{2} = \frac{-8-3}{2} = \frac{-11}{2} = -5.5$

At the point (2,3), the tangent slope to the curve $2x^2 + xy = 2\ is - 5.5$

6. Incorrect

At point (3,4) of the curve $2x^2 + 3xy - y^2 = 38$ the slope of the tangent line is: $= -1.41$

Using implicit differentiation, we have:

$(2x^2)' + (3xy)' + (y^2)' = 0$

$4x + 3[(x)'y + x(y)'] + 2y(y)' = 0$

$4x + 3(y + xy') + 2y(y)' = 0$

$4x + 3y + 3xy' + 2y(y)' = 0$

$3xy' + 2y(y)' = -4x - 3y$

$y'(3x + 2y) = -4x - 3y$

$y' = \frac{-4x-3y}{3x+2y}$

The slope of the tangent at point (3,4) is: $y' = \frac{-4(3)-3(4)}{3(3)+2(4)} = \frac{-12-12}{9+8} = \frac{-24}{17} = -1.41$

7. Correct

The slope of the tangent line to the graph of $y = \frac{4}{\pi}x - \sin(xy)\ at\ \left(\frac{\pi}{2}, 1\right)\ is\ \frac{4}{\pi}$

Using implicit differentiation, we have:

$y' = (\frac{4}{\pi}x)' - [\sin(xy)]' = \frac{4}{\pi} - \cos(xy)(xy)' = \frac{4}{\pi} - \cos(xy)(y + xy')$

We are replacing $x = \frac{\pi}{2}$ and y=1 in

$y' = \frac{4}{\pi} - \cos(xy)(y + xy')$

So,

$y' = \frac{4}{\pi} - \cos\left(\frac{\pi}{2} * 1\right)\left(1 + \frac{\pi}{2}y'\right); but\ \cos\left(\frac{\pi}{2}\right) = 0\ then\ y' = \frac{4}{\pi}$

8. Incorrect

If $3y^2 + \ln(x) = 2y - \cos(x)$, $y' = \frac{\sin(x) - \frac{1}{x}}{6y-2}$

Using implicit differentiation, we have:

$(3y^2)' + [\ln(x)]' = (2y)' - [\cos(x)]'$

$6y(y)' + \frac{1}{x} = 2y' + \sin(x)$

$6y(y)' - 2y' = +\sin(x) - \frac{1}{x}$

$y'(6y - 2) = \sin(x) - \frac{1}{x}$

$y' = \frac{\sin(x) - \frac{1}{x}}{6y-2}$

9. Incorrect

If $\cot(y) + 2x = 5y - y^2$, $y' = \frac{-2}{-csc^2(y) - 5 + 2y}$

$[\cot(y)]' + (2x)' = (5y)' - (y^2)'$

$-csc^2(y)(y)' + 2 = 5y' - 2y\,y'$

$-csc^2(y)(y)' - 5y' + 2y\,y' = -2$

$y'[-csc^2(y) - 5 + 2y] = -2$

$y' = \frac{-2}{-csc^2(y)-5\ +2y}$

10. Correct

If $(y-1)^2 = 6y + x^3 + 2x,\ y' = \frac{3x^2+2}{2(y-1)-6}$

Using implicit differentiation, we have:

$[(y-1)^2]' = (6y)' + (x^3)' + (2x)'$

$2(y-1)(y-1)' = 6y' + 3x^2 + 2$

$2(y-1)y' = 6y' + 3x^2 + 2$

$2(y-1)y' - 6y' = 3x^2 + 2$

$y'[2(y-1) - 6] = 3x^2 + 2$

$y' = \frac{3x^2+2}{2(y-1)-6}$

Chapter 3. I. a. Relating graph of f(x) to f '(x) and f ''(x)

1. Correct

If the first derivative is positive, the function is increasing.

2. Correct

If the second derivative is positive, the function is concave up.

3. incorrect

The graph of the function $f(x) = \frac{2}{3}x^3 - 2x^2 - 6x + 7$ is decreasing between -1 and 3

$f'(x) = (\frac{2}{3}x^3)' - (2x^2)' - (6x)' + (7)' = \frac{6}{3}x^2 - 4x - 6 = 2x^2 - 4x - 6 = 2(x^2 - 2x - 3) = 2(x + 1)(x - 3)$

$f'(x) = 2(x+1)(x-3)$ is zero for x=-1 and x=3

We check the sign of the first derivative by considering x values less than -1, between -1 and 3, and bigger than 3.

$f'(-2) = 2[(-2)+1][(-2)-3] = 2(-1)(-5) = +10$

The first derivative is positive for values of x less than -1

$f'(0) = 2[(0)+1][(0)-3] = 2(1)(-3) = -6$

The first derivative is negative between -1 and 3

$f'(4) = 2[(4)+1][(4)-3] = 2(5)(1) = +10$

The first derivative is positive for values of x bigger than 3

The sign diagram for $f'(x)$ is shown below:

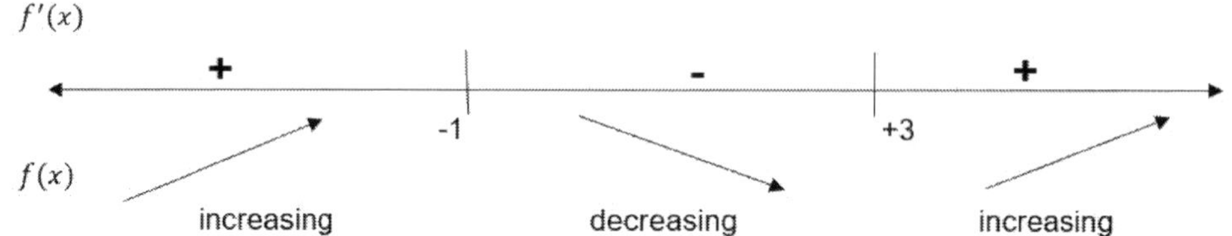

The graph is shown below.

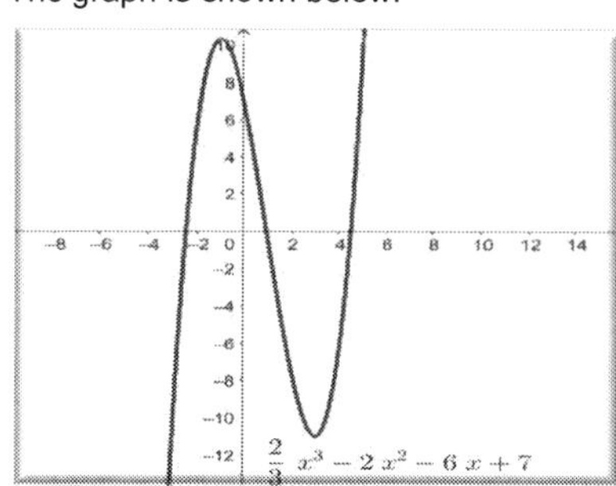

4. Correct

The graph of the function $f(x) = 2x^3 - 7x^2 + 7$ is decreasing between 0 and 2.33

$f'(x) = 6x^2 - 14x = 2x(3x - 7)$

$f'(x) = 2x(3x - 7) = 0$ for x=0 and x=7/3=2.33

$f'(-1) = 2(-1)[3(-1) - 7] = -2(-10) = +20$

The first derivative is positive for values of x less than zero.

$f'(1) = 2(1)[3(1) - 7] = 2(-4) = -8$

The first derivative is negative for values of x between 0 and 2.33

$f'(3) = 2(3)[3(3) - 7] = 6(2) = +12$

The first derivative is positive for values of x bigger than 2.33

The sign of the first derivative and the behavior of the graph is shown below

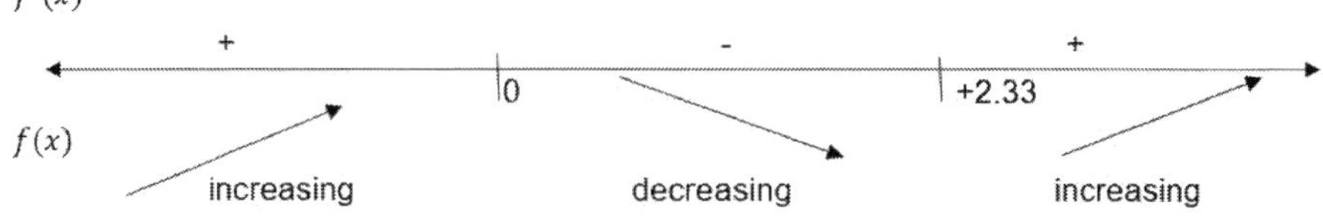

The graph of the function is shown below.

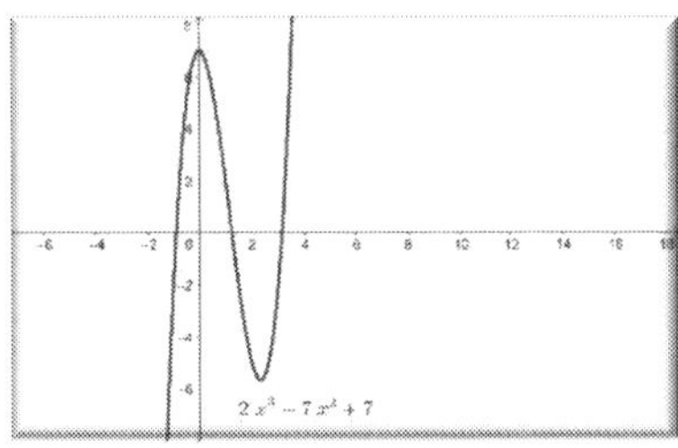

5. Correct

The graph of the function $f(x) = \frac{2x-5}{x-1}$ is concave up for values of x less than 1

First, the graph of the function will intersect x axis at x=2.5

Second, there is a vertical asymptote at x=1

$\lim_{x\to 1^-} \frac{2x-5}{x-1} = +\infty$

$\lim_{x\to 1^+} \frac{2x-5}{x-1} = -\infty$

Third, $\lim_{x\to\infty} \frac{2x-5}{x-1} = 2 = \lim_{x\to-\infty} \frac{2x-5}{x-1}$. The graph will have a horizontal asymptote at y=2

Using the quotient rule, we have:

$f'(x) = \frac{(2x-5)'(x-1)-(2x-5)(x-1)'}{(x-1)^2} = \frac{2(x-1)-(2x-5)}{(x-1)^2} = \frac{2x-2-2x+5}{(x-1)^2} = \frac{3}{(x-1)^2}$ is always positive

To verify the concavity, we have to calculate the second derivate.

$f''(x) = [3(x-1)^{-2}]' = 3(-2)(x-1)^{-3}(x-1)' = -6(x-1)^{-3} = \frac{-6}{(x-1)^3}$

$f''(0) = 6, and\ f''(2) = -6$

The sign and concavity of the graph of the function $f(x) = \frac{2x-5}{x-1}$ is shown below.

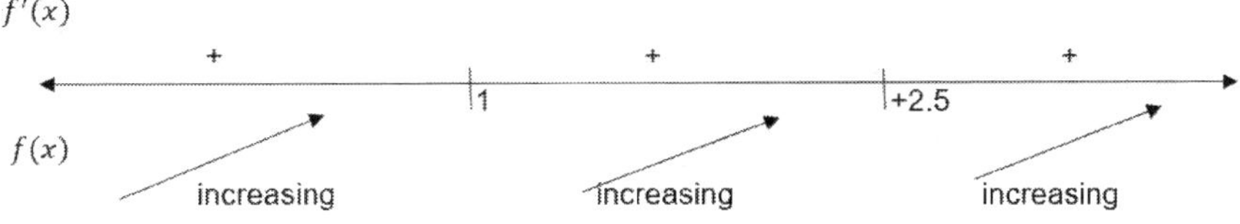

The concavity is shown below

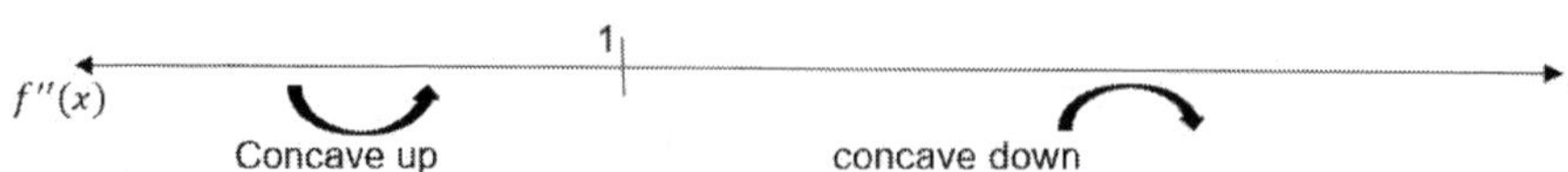

The graph is shown below.

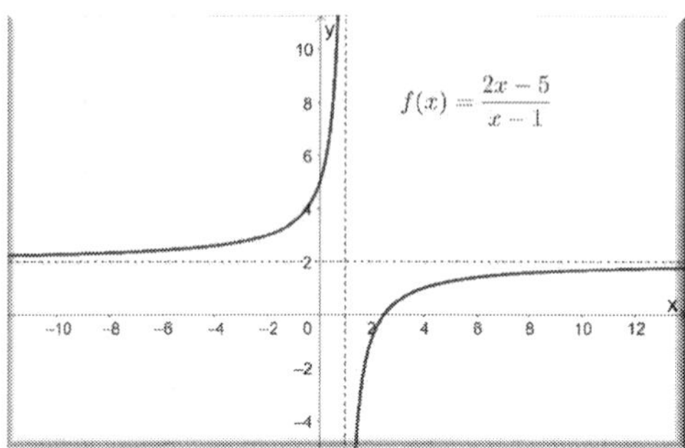

6. Incorrect

The function $f(x) = \frac{x^2}{x^2-4}$ is concave down between -2 and 2

$f(x) = \frac{x^2}{x^2-4} = \frac{x^2}{(x-2)(x+2)}$ 0

First, the graph of the function will intersect x axis at x=0, because $x^2 = 0$ for x=0

Second, there are two vertical asymptotes at x=-2 and x=2

The sign of the function is shown below

	-2		0	2
x^2	+	+	+	+
$x^2 - 4$	+	-	-	+
Sign of function	+	-	-	+

$\lim_{x \to -2^-} \frac{x^2}{x^2-4} = +\infty$

$\lim_{x \to -2^+} \frac{x^2}{x^2-4} = -\infty$

$\lim_{x \to 2^-} \frac{x^2}{x^2-4} = -\infty$

$\lim_{x \to 2^+} \frac{x^2}{x^2-4} = +\infty$

Third, $\lim_{x \to \infty} \frac{x^2}{x^2-4} = 1 = \lim_{x \to -\infty} \frac{x^2}{x^2-4}$. The graph will have a horizontal asymptote at y=1

Using the quotient rule, we have:

$$f'(x) = \frac{(x^2)'(x^2-4)-(x^2)(x^2-4)'}{(x^2-4)^2} = \frac{(2x)(x^2-4)-(x^2)(2x)}{(x^2-4)^2} = \frac{2x^3-8x-2x^3}{(x^2-4)^2} = \frac{-8x}{(x^2-4)^2}$$

$f'(x)$ is zero at x=0.

The sign of the first derivative is shown below.

	-2		0	2
$-8x$	+	+	-	-
$f'(x)$	+	+	-	-
Graph	↗	↗	↘	↘

To verify the concavity, we have to calculate the second derivate. We are using the product rule.

$$f''(x) = [-8x(x^2-4)^{-2}]' = (-8x)'(x^2-4)^{-2} + (-8x)[(x^2-4)^{-2}]' = -8(x^2-4)^{-2} - 8x(-2)(x^2-4)^{-3}(2x) = -8(x^2-4)^{-2} + 32x^2(x^2-4)^{-3} = \frac{32x^2}{(x^2-4)^3} - \frac{8}{(x^2-4)^2} = \frac{32x^2}{(x^2-4)^3} - \frac{8(x^2-4)}{(x^2-4)^3} =$$

$$\frac{32x^2-8(x^2-4)}{(x^2-4)^3} = \frac{32x^2-8x^2+32}{(x^2-4)^3} = \frac{24x^2+32}{(x^2-4)^3}$$

$24x^2+32 = 0\ so\ x = \sqrt{\frac{-32}{24}}$ not a real number.

$24x^2+32$ is positive for any x values.

For x=-3

$[(-3)^2-4]^3 = [9-4]^3 = 5*5*5 = 125$ positive

For x=0

$[(0)^2-4]^3 = [0-4]^3 = (-4)*(-4)*(-4) = -64$ negative

For x=3

$[(3)^2-4]^3 = [9-4]^3 = 5*5*5 = 125$ positive

The concavity is shown below

	-2	2	
$(x^2-4)^3$	+	-	+
$f''(x)$	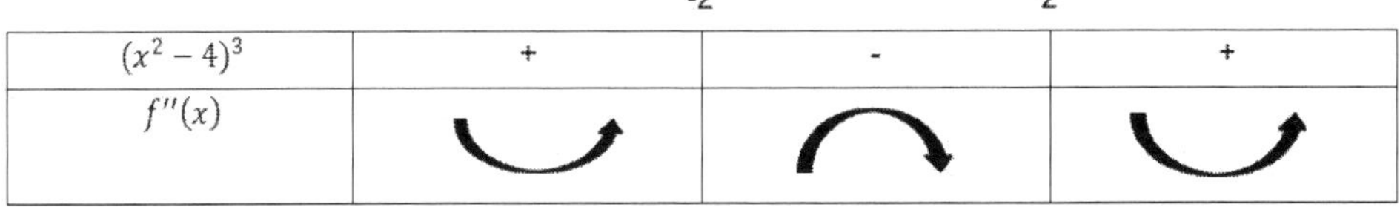		

The graph is represented below

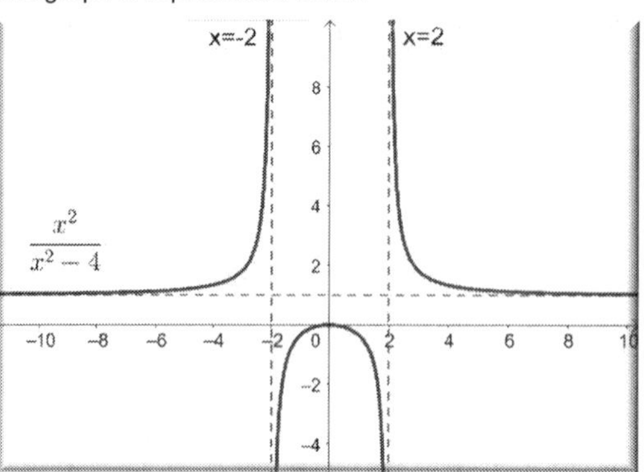

7. Incorrect

The function $f(x) = \frac{2x}{x+3}$ is concave down for x bigger than -3.

First, the graph of the function will intersect x axis at x=0.

Second, there is one vertical asymptote at x=-3.

	-3	0	
$2x$	-	-	+
$x+3$	-	+	+
Sign of function	+	-	+

$\lim\limits_{x \to -3^-} \frac{2x}{x+3} = +\infty$

$\lim\limits_{x \to -3^+} \frac{2x}{x+3} = -\infty$

Third, $\lim\limits_{x \to \infty} \frac{2x}{x+3} = 2 = \lim\limits_{x \to -\infty} \frac{2x}{x+3}$. The graph will have a horizontal asymptote, at y=2

Using the quotient rule, we have:

$f'(x) = \frac{(2x)'(x+3)-(2x)(x+3)'}{(x+3)^2} = \frac{2(x+3)-2x}{(x+3)^2} = \frac{2x+6-2x}{(x+3)^2} = \frac{6}{(x+3)^2}$; $\frac{6}{(x+3)^2}$ is always positive.

The sign of the first derivative is shown below.

$f'(x)$	+
Graph	↗

To verify the concavity, we have to calculate the second derivate.

$f''(x) = \frac{6}{(x+3)^2} = 6(x+3)^{-2} = 6(-2)(x+3)^{-3} = \frac{-12}{(x+3)^3}$

The concavity results from the graph below.

		-3
-12	-	-
$(x+3)^3$	-	+
$f''(x)$	+	-
Concavity	∪	∩

The graph is shown below.

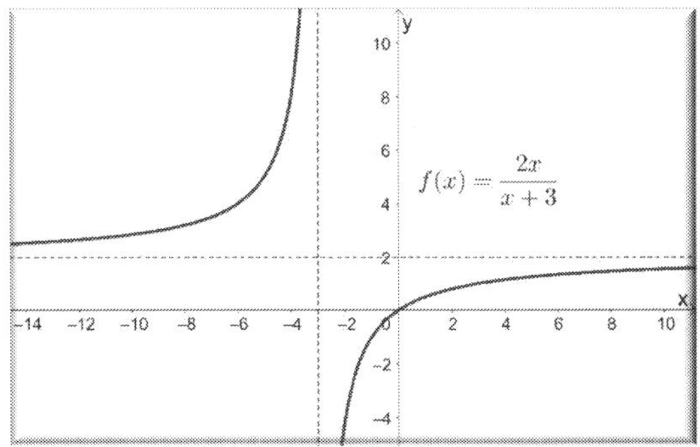

8. Correct

The function $f(x) = \frac{1}{4x^2-9}$ is concave down between x=-1.5 and x=1.5

First, the graph of the function will never intersect x axis.

$f(x) = \frac{1}{4x^2-9} = \frac{1}{(2x-3)(2x+3)}$

Second, there are two vertical asymptotes at x=-1.5and x=1.5

		-1.5	1.5
$2x-3$	-	-	+
$2x+3$	-	+	+
Sign of function	+	-	+

$\lim_{x\to -1.5^-} \frac{1}{4x^2-9} = +\infty$

$\lim_{x\to -1.5^+} \frac{1}{4x^2-9} = -\infty$

$\lim_{x\to 1.5^-} \frac{1}{4x^2-9} = -\infty$

$\lim_{x\to -1.5^+} \frac{1}{4x^2-9} = +\infty$

Third, $\lim_{x\to\infty} \frac{1}{4x^2-9} = 0 = \lim_{x\to-\infty} \frac{1}{4x^2-9}$. The graph will have a horizontal asymptote, at y=0

We have:

$f'(x) = [(4x^2-9)^{-1}]' = -(4x^2-9)^{-2}(4x^2-9)' = -(4x^2-9)^{-2}(8x) = \frac{-8x}{(4x^2-9)^2}$ is zero at x=0

The sign of the first derivative is shown below.

	0	
$-8x$	+	-
$(4x^2-9)^2$	+	+
$f'(x)$	-	+
Graph		

To verify the concavity, we have to calculate the second derivate. We use the product rule.

$f''(x) = [\frac{-8x}{(4x^2-9)^2}]' = [-8x(4x^2-9)^{-2}]' = (-8x)'(4x^2-9)^{-2} + (-8x)[(4x^2-9)^{-2}]' =$

$-8(4x^2-9)^{-2} - 8x(-2)(4x^2-9)^{-3}(4x^2-9)' = -8(4x^2-9)^{-2} + 8x(2)(4x^2-9)^{-3}(8x) = \frac{-8}{(4x^2-9)^2} +$

$\frac{128x^2}{(4x^2-9)^3} = \frac{-8(4x^2-9)+128x^2}{(4x^2-9)^3} = \frac{-32x^2+72+128x^2}{(4x^2-9)^3} = \frac{96x^2+72}{(4x^2-9)^3} = \frac{24(4x^2+3)}{(4x^2-9)^3}$

The concavity results from the graph below.

	-1.5	1.5	
$24(4x^2+3)$	+	+	+
$(4x^2-9)^3$	+	-	+
$f''(x)$	+	-	+
Concavity	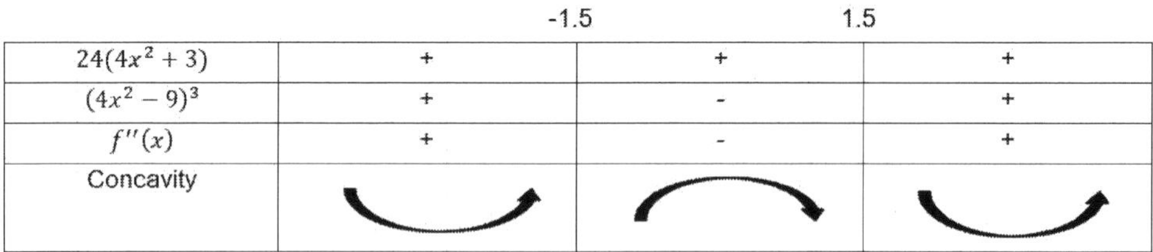		

The graph is shown below.

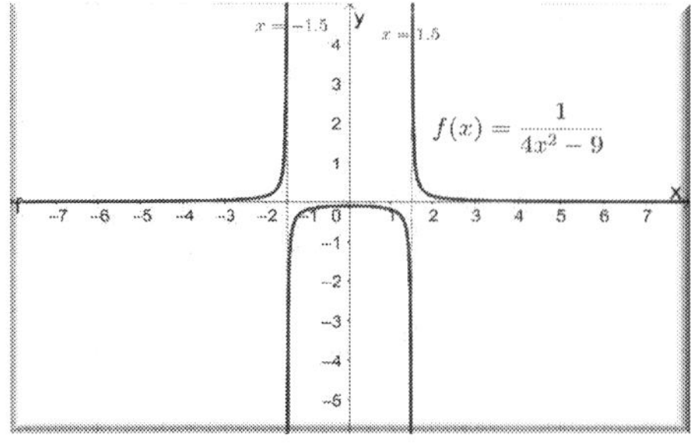

9. Incorrect

The function $f(x) = \frac{x}{x^2-4}$ is concave up between x=-2 and x=0 and x>2

First, the graph of the function will intersect x axis at =0.

$f(x) = \frac{x}{x^2-4} = \frac{x}{(x-2)(x+2)}$

Second, there are two vertical asymptotes at x=-2 and x=2

		-2	0	2
x	-	-	+	+
$x-2$	-	-	-	+
$x+2$	-	+	+	+
Sign of function	-	+	-	+

$\lim_{x\to-2^-} \frac{x}{x^2-4} = -\infty$

$\lim_{x\to-2^+} \frac{x}{x^2-4} = +\infty$

$\lim_{x\to2^-} \frac{x}{x^2-4} = -\infty$

$\lim_{x\to2^+} \frac{x}{x^2-4} = +\infty$

Third, $\lim_{x\to\infty} \frac{x}{x^2-4} = 0 = \lim_{x\to-\infty} \frac{x}{x^2-4}$. The graph will have a horizontal asymptote, at y=0

We have:

$f'(x) = [x(x^2-4)^{-1}]' = x'(x^2-4)^{-1} + x[(x^2-4)^{-1}]' = (x^2-4)^{-1} + x(-1)(x^2-4)^{-2}(x^2-4)' =$
$(x^2-4)^{-1} - x(x^2-4)^{-2}(2x) = \frac{1}{x^2-4} - \frac{2x^2}{(x^2-4)^2} = \frac{x^2-4}{(x^2-4)^2} - \frac{2x^2}{(x^2-4)^2} = \frac{x^2-4-2x^2}{(x^2-4)^2} = \frac{-x^2-4}{(x^2-4)^2}$ is never zero.

The sign of the first derivative is shown below.

$-x^2-4$	-
$(x^2-4)^2$	+
$f'(x)$	-
Graph	↘

To verify the concavity, we have to calculate the second derivate.

$f''(x) = [\frac{-x^2-4}{(x^2-4)^2}]' = [(-x^2-4)(x^2-4)^{-2}]' = (-x^2-4)'(x^2-4)^{-2}] + (-x^2-4)[(x^2-4)^{-2}]' =$
$-2x(x^2-4)^{-2} + (-x^2-4)(-2)(x^2-4)^{-3}(2x) = \frac{-2x}{(x^2-4)^2} + \frac{4x(x^2+4)}{(x^2-4)^3} = \frac{-2x(x^2-4)}{(x^2-4)^3} + \frac{4x(x^2+4)}{(x^2-4)^3} =$
$\frac{-2x^3+8x+4x^3+16x}{(x^2-4)^3} = \frac{2x^3+24x}{(x^2-4)^3} = \frac{2x(x^2+12)}{(x^2-4)^3}$

The concavity is shown below

	-2	0	2	
x	-	-	+	+
x^2+12	+	+	+	+
$(x^2-4)^3$	+	-	-	+
$f''(x)$	-	+	-	+
Concavity	⌒	∪	⌒	∪

The graph is shown below.

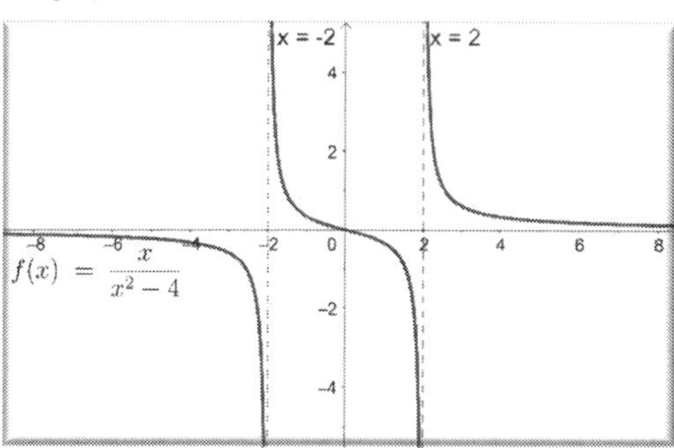

10. Correct

The function $f(x) = x^5 - 2x^3$ is concave up for $-0.77 \le x \le 0$ *and* $0.77 \le x \le \sqrt{2}$

First, the graph of the function will intersect x axis at x=0 ;, x=$-\sqrt{2}$ and, x=$\sqrt{2}$

$f(x) = x^3(x^2-2)$

Second, there are no asymptotes.

	$-\sqrt{2}$	0	$\sqrt{2}$	
x^3	-	-	+	+
$x-\sqrt{2}$	-	-	-	+
$x+\sqrt{2}$	-	+	+	+
Sign of fuction	-	+	-	+

Third,

$\lim_{x\to\infty}(x^5-2x^3) = +\infty.$

$\lim_{x\to-\infty}(x^5-2x^3) = -\infty$

We have:

$f'(x) = (x^5 - 2x^3)' = 5x^4 - 6x^2 = x^2(5x^2 - 6)$ is zero at x=0, x=-1.09 and, x=1.09

The sign of the first derivative is shown below.

		-1.09	0	1.09
x^2	+	+	+	+
$5x^2 - 6$	+	-	-	+
$f'(x)$	+	-	-	+
Graph	↗	↘	↘	↗

To verify the concavity, we have to calculate the second derivate.

$f''(x) = [x^2(5x^2 - 6)]' = (x^2)'(5x^2 - 6) + x^2(5x^2 - 6)' = 2x(5x^2 - 6) + x^2(10x) = 10x^3 - 12x + 10x^3 = 4x(5x^2 - 3)$ is zero at x=0, x=-0.77 and x=0.77

		$-\sqrt{2}$	-0.77	0	0.77	$\sqrt{2}$
$4x$	-	-	-	+	+	+
$5x^2 - 3$	+	+	-	-	+	+
Sign of $f''(x)$	-	-	+	-	+	+
Concavity	⌒	⌒	∪	⌒	∪	∪

The graph of the function $f(x) = x^5 - 2x^3$ is shown below.

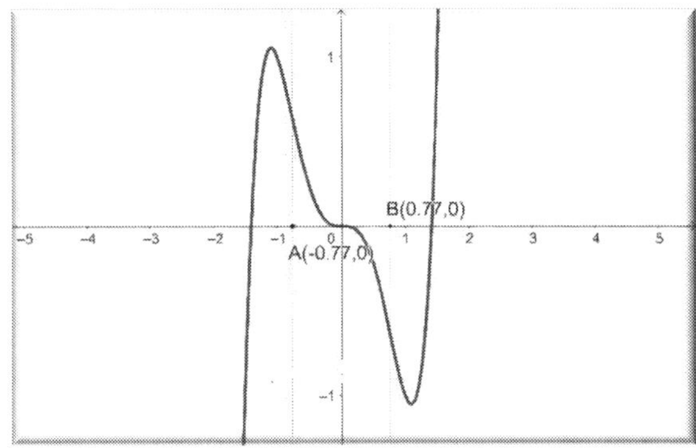

Chapter 3. I. b. Differentiability, mean value theorem

1. Correct

A differentiable function is a function whose derivative exists all over the domain.

2. Correct

In a planar arc between two points, there is at least one point at which the tangent to the arc is parallel to the secant through the arc's two points.

3. Incorrect

If $f(x) = 3x^2 - 4x + 1$ and the points (-3,40) and, (1,0) the value of x where the tangent at the graph is parallel with the line that goes through (-3,40) and, (1,0) is: x=-1

$f(-3) = 3(-3)^2 - 4(-3) + 1 = 27 + 12 + 1 = 40$

$f(1) = 3(1)^2 - 4(1) + 1 = 3 - 4 + 1 = 0$

We are calculating the slope of the line that passes through (-3,40) and, (1,0)

$slope = \frac{f(1)-f(-3)}{1-(-3)} = \frac{0-40}{1+3} = \frac{-40}{4} = -10$

We calculate the first derivative.

$f'(x) = (3x^2 - 4x + 1)' = 6x - 4$

We equalize $f'(x)$ with the value of the slope, in this case -10

$f'(x) = -10$

$f'(x) = 6x - 4 = -10, so\ x = \frac{-10+4}{6} = -1$

4. Incorrect

If $f(x) = x^2 + 5x - 3$ and the points (-1,-7) and, (2,11) the value of x where the tangent at the graph is parallel with the line that goes through (-1,-7) and, (2,11) is: x=$\frac{1}{2}$

$f(-1) = (-1)^2 + 5(-1) - 3 = 1 - 5 - 3 = -7$

$f(2) = (2)^2 + 5(2) - 3 = 4 + 10 - 3 = 11$

We are calculating the slope of the line that passes through (-3,40) and, (1,0)

$slope = \frac{f(2)-f(-1)}{2-(-1)} = \frac{11-(-7)}{2+1} = \frac{18}{3} = 6$

We calculate the first derivative.

$f'(x) = (x^2 + 5x - 3)' = 2x + 5$

We equalize $f'(x)$ with the value of the slope, in this case 6

$f'(x) = 6$

$f'(x) = 2x + 5 = 6, so\ x = \frac{6-5}{2} = \frac{1}{2}$

5. Correct

If $f(x) = x^5 - 2x^3$and the points (-1.6,-2.28) and, (1.6,2.28) the values of x where the tangents at the graph are parallel with the line that goes through (-1.6,-2.28) and, (1.6,2.28) are: x=∓1.18

$f(-1.6) = (-1.6)^5 - 2(-1.6)^3 = -10.48 - 2(-4.096) = -2.28$

$f(1.6) = (1.6)^5 - 2(1.6)^3 = 10.48 - 2(4.096) = 2.28$

$slope = \frac{f(1.6)-f(-1.6)}{1.6-(-1.6)} = \frac{2.28-(-2.28)}{3.2} = \frac{4.57}{3.2} = 1.43$

$f'(x) = (x^5 - 2x^3)' = 5x^4 - 6x^2 = x^2(5x^2 - 6) = 1.43$

If we substitute x^2 with A, we have:

$f'(x) = A(5A - 6) = 1.43$

$5A^2 - 6A - 1.43 = 0$

$A_{1,2} = \frac{6\mp\sqrt{36-4(5)(-1.43)}}{2(5)} = \frac{6\mp\sqrt{64.6}}{10} = \frac{6\mp 8.03}{10}$

$A_1 = \frac{6-8.03}{10} = -0.203$ It will give non real x values.

$A_2 = \frac{6+8.03}{10} = 1.403$

$x^2 = 1.403, so\ x = \mp 1.18$

6. Incorrect

If $f(x) = 4x^3 + 3x^2$ the values of x where the tangents of 1 at the graph are: x=-0.13 and x=0.63

$f'(x) = (4x^3 - 3x^2)' = 12x^2 - 6x = 1$

$12x^2 - 6x - 1 = 0$

$x_{1,2} = \frac{6\mp\ \sqrt{36-4(12)(-1)}}{2*12} = \frac{6\mp\sqrt{84}}{24} = \frac{6\mp 9.16}{24}$

$x_1 = \frac{6+9.19}{24} = \frac{15.19}{24} = 0.63$

$x_2 = \frac{6-9.19}{24} = \frac{-3.19}{24} = -0.13$

7. Incorrect

If $f(x) = 6x^4 + 7$ the values of x where the tangent of -3 at the graph is: $x = -\frac{1}{2}$

$f'(x) = (6x^4 + 7)' = 24x^3$

$24x^3 = -3, so\ x = \sqrt[3]{\frac{-3}{24}} = \sqrt[3]{\frac{-1}{8}} = -\frac{1}{2}$

8. Correct

If $f(x) = \frac{1}{x^2} - \frac{2}{5}$ the value of x where the tangent of 2 at the graph is: x=-1

$f'(x) = (\frac{1}{x^2} - \frac{2}{5})' = (x^{-2} - 0.4)' = -2x^{-3} = -\frac{2}{x^3} = 2$

$f'(x) = -\frac{1}{x^3} = 1, so - 1 = x^3$

$x^3 = -1$

$x = -1$

9. Correct

If $f(x) = \ln(x) + 10$ the value of x where the tangent of 5 at the graph is: x= $\frac{1}{5}$

$f'(x) = [\ln(x) - \frac{2}{5}]' = \frac{1}{x}$

$\frac{1}{x} = 5, so\ x = \frac{1}{5}$

10. Incorrect

If $f(x) = x^3 + 2$ and the points (-0.5,1.875) and, (0.5,2.125) the values of x where the tangents at the graph are parallel with the line that goes through (-0.5,1.875) and, (0.5,2.125) are: x=∓0.28

$f(-0.5) = (-0.5)^3 + 2 = -0.125 + 2 = 1.875$

$f(0.5) = (0.5)^3 + 2 = 0.125 + 2 = 2.125$

$\frac{f(0.5)-f(-0.5)}{0.5-(-0.5)} = \frac{2.125-1.875}{1} = 0.25$

$f'(x) = 3x^2 = 0.25$

$x^2 = \frac{0.25}{3}\ so, x = \mp\sqrt{\frac{0.25}{3}} = \mp 0.28$

Chapter 3. I. c. Newton's method

1. Correct

The square root of 135, using Newton's method is: 11.62

$x^2 = 135$

$f(x) = x^2 - 135$

$f'(x) = 2x$

We decide the initial value as 5.

$f(5) = 5^2 - 135 = 25 - 135 = -110$

$x_1 = x_0 - \frac{f(x_0)}{f'(x_0)}$

$x_1 = 5 - \frac{f(x_0)}{f'(x_0)} = 5 - \frac{-110}{2(5)} = 5 + \frac{110}{10} = 5 + 11 = 16$

$x_2 = 16 - \frac{f(16)}{f'(16)} = 16 - \frac{16^2-135}{2(16)} = 16 - \frac{256-135}{32} = 16 - 3.78 = 12.21$

$x_3 = 12.21 - \frac{f(12.21)}{f'(12.21)} = 12.21 - \frac{12.21^2-135}{2(12.21)} = 12.21 - \frac{149.08-135}{24.42} = 12.21 - 0.57 = 11.63$

$x_4 = 11.63 - \frac{f(11.63)}{f'(11.63)} = 11.63 - \frac{11.63^2-135}{2(11.63)} = 11.63 - \frac{135.29-135}{23.26} = 11.63 - 0.01 = 11.62$

2. Incorrect

The square root of 432, using Newton's method is: 25.43

$x^2 = 432$

$f(x) = x^2 - 432$

$f'(x) = 2x$

We decide the initial value as 15.

$f(15) = 15^2 - 432 = 225 - 432 = -207$

$x_1 = x_0 - \frac{f(x_0)}{f'(x_0)}$

$x_1 = 15 - \frac{f(x_0)}{f'(x_0)} = 15 - \frac{-207}{2(15)} = 15 + \frac{207}{30} = 15 + 69 = 81$

$x_2 = 81 - \frac{f(81)}{f'(81)} = 81 - \frac{81^2-432}{2(81)} = 81 - \frac{6561-432}{162} = 81 - 37.83 = 43.16$

$x_3 = 43.16 - \frac{f(43.16)}{f'(43.16)} = 43.16 - \frac{43.16^2-432}{2(43.16)} = 43.16 - \frac{1863-432}{86.32} = 43.16 - 16.57 = 26.58$

$x_4 = 26.58 - \frac{f(26.58)}{f'(26.58)} = 26.58 - \frac{26.58^2-432}{2(26.58)} = 26.58 - \frac{706.49-432}{53.16} = 26.58 - 5.16 = 21.41$

$x_5 = 21.41 - \frac{f(21.41)}{f'(21.41)} = 21.41 - \frac{21.41^2-432}{2(21.41)} = 21.41 - \frac{458.38-432}{42.82} = 21.41 - 0.61 = 20.79$

$x_5 = 20.79 - \frac{f(20.79)}{f'(20.79)} = 20.79 - \frac{20.79^2-432}{2(20.79)} = 20.79 - \frac{432.22-432}{41.58} = 20.79 - 0.005 = 20.78$

3. Incorrect

The solution of $x^3 - 2x^2 = 4$ is: 5

$f(x) = x^3 - 2x^2 - 4$

$f'(x) = 3x^2 - 4x$

$x_1 = x_0 - \frac{f(x_0)}{f'(x_0)}$

We decide the initial value as 2.

$x_1 = 2 - \frac{2^3-2(2)^2-4}{3(2)^2-4(2)} = 2 - \frac{8-8-4}{12-8} = 2 + 1 = 3$

$x_2 = 3 - \frac{3^3-2(3)^2-4}{3(3)^2-4(3)} = 3 - \frac{5}{15} = 3 - 0.33 = 2.66$

$x_3 = 2.66 - \frac{2.66^3-2(2.66)^2-4}{3(2.66)^2-4(2.66)} = 2.66 - \frac{0.74}{10.66} = 2.66 - 0.069 = 2.59$

4. Correct

The solution of $3x^4 - 4x^3 + 2x = 7$ is: x=1.624

$f(x) = 3x^4 - 4x^3 + 2x - 7$

$f'(x) = 12x^3 - 12x^2 + 2$

$x_1 = x_0 - \frac{f(x_0)}{f'(x_0)}$

We decide the initial value as 2.

$x_1 = 2 - \frac{3(2)^4-4(2)^3+2(2)-7}{12(2)^3-12(2)^2+2} = 2 - \frac{3(16)-4(8)+2(2)-7}{12(8)-12(4)+2} = 2 - \frac{13}{50} = 2 - 0.26 = 1.74$

$x_2 = 1.74 - \frac{3(1.74)^4-4(1.74)^3+2(1.74)-7}{12(1.74)^3-12(1.74)^2+2} = 1.74 - \frac{2.9}{28.88} = 1.74 - 0.10064 = 1.63$

$x_3 = 1.63 - \frac{3(1.63)^4-4(1.63)^3+2(1.63)-7}{12(1.63)^3-12(1.63)^2+2} = 1.63 - \frac{0.32}{22.61} = 1.63 - 0.014 = 1.625$

$x_4 = 1.625 - \frac{3(1.625)^4-4(1.625)^3+2(1.625)-7}{12(1.625)^3-12(1.625)^2+2} = 1.625 - \frac{0.0058}{21.80} = 1.625 - 0.000267 = 1.624$

5. Incorrect

The solution of $2x^3+3x-4=\sin(x)$ is: x=0.980

$f(x)=2x^3+3x-4-\sin(x)$

$f'(x)=6x^2+3-\cos(x)$

$x_1=x_0-\frac{f(x_0)}{f'(x_0)}$

We decide the initial value as -1.

$x_1=-1-\frac{2(-1)^3+3(-1)-\sin(-1)}{6(-1)^2+3-\cos(-1)}=-1+\frac{8.15}{5.45}=-1+1.49=0.49$

$x_2=0.49-\frac{2(0.49)^3+3(0.49)-\sin(0.49)}{6(0.49)^2+3-\cos(0.49)}=0.49-\frac{-2.74}{0.58}=0.49+4.69=5.18$

$x_3=5.18-\frac{2(5.18)^3+3(5.18)-\sin(5.18)}{6(5.18)^2+3-\cos(5.18)}=5.18-\frac{291.8}{161.06}=5.18-1.81=3.37$

$x_4=3.37-\frac{2(3.37)^3+3(3.37)-\sin(3.37)}{6(3.37)^2+3-\cos(3.37)}=3.37-\frac{83.36}{69.38}=3.37-1.2=2.17$

$x_5=2.17-\frac{2(2.17)^3+3(2.17)-\sin(2.17)}{6(2.17)^2+3-\cos(2.17)}=3.37-\frac{22.28}{28.95}=3.37-0.76=1.4$

$x_6=1.4-\frac{2(1.4)^3+3(1.4)-\sin(1.4)}{6(1.4)^2+3-\cos(1.4)}=1.4-\frac{4.78}{11.68}=1.4-0.4=0.99$

$x_7=0.99-\frac{2(0.99)^3+3(0.99)-\sin(0.99)}{6(0.99)^2+3-\cos(0.99)}=0.99-\frac{1.12}{5.41}=0.99-0.023=0.97$

$x_8=0.97-\frac{2(0.97)^3+3(0.97)-\sin(0.97)}{6(0.97)^2+3-\cos(0.97)}=0.97+\frac{0.06}{5.114}=0.97+0.013=0.985$

$x_9=0.985-\frac{2(0.985)^3+3(0.985)-\sin(0.985)}{6(0.985)^2+3-\cos(0.985)}=0.985-\frac{0.08}{5.27}=0.985-0.007=0.978$

$x_{10}=0.978-\frac{2(0.978)^3+3(0.978)-\sin(0.978)}{6(0.978)^2+3-\cos(0.978)}=0.978+\frac{0.02}{5.18}=0.978+0.004=0.982$

$x_{11}=0.982-\frac{2(0.982)^3+3(0.982)-\sin(0.982)}{6(0.982)^2+3-\cos(0.982)}=0.982-\frac{0.013}{5.23}=0.982-0.002=0.980$

6. Correct

The solution of $x^5+4x^3-5x=3$ is: x=1.18

$f(x)=x^5+4x^3-5x-3$

$f'(x)=5x^4+12x^2-5$

$x_1=x_0-\frac{f(x_0)}{f'(x_0)}$

We decide the initial value as 2.

$x_1=2-\frac{2^5+4(2)^3-5(2)-3}{5(2)^4+12(2)^2-5}=2-\frac{51}{123}=2-0.41=1.585$

$x_2=1.585-\frac{(1.585)^5+4(1.585)^3-5(1.585)-3}{5(1.585)^4+12(1.585)^2-5}=1.585-\frac{15.02}{56.74}=1.585-0.26=1.32$

$x_3=1.32-\frac{(1.32)^5+4(1.32)^3-5(1.32)-3}{5(1.32)^4+12(1.32)^2-5}=1.32-\frac{3.62}{31.13}=1.32-0.11=1.2$

$x_4=1.2-\frac{(1.2)^5+4(1.2)^3-5(1.2)-3}{5(1.2)^4+12(1.2)^2-5}=1.2-\frac{0.49}{22.91}=1.2-0.021=1.182$

$x_4 = 1.182 \; - \frac{(1.182\;)^5+4(1.182\;)^3-5(1.182\;)-3}{5(1.182\;)^4+12(1.182\;)^2-5} = 1.182 \; - \frac{0.014}{21.55} = 1.182 \; - 0.0006 = 1.1817$

7. Correct

The solution of $(x-3)^3 = \sin(x)$ is: x=3.138

$f(x) = (x-3)^3 - \sin(x)$

$f'(x) = 3(x-3)^2 - \cos(x)$

$x_1 = x_0 - \frac{f(x_0)}{f'(x_0)}$

We decide the initial value as 2.

$x_1 = 2 - \frac{(2-3)^3-\sin(2)}{3(2-3)^2-\cos(2)} = 2 + \frac{1.9}{3.41} = 2 + 0.558 = 2.558$

$x_2 = 2.558 - \frac{(2.558-3)^3-\sin(2.558)}{3(2.558-3)^2-\cos(2.558)} = 2.558 + \frac{0.63}{1.41} = 2.558 + 0.448 = 3.007$

$x_3 = 3.007 - \frac{(3.007-3)^3-\sin(3.007)}{3(3.007-3)^2-\cos(3.007)} = 3.007 + \frac{0.133}{0.99} = 3.007 + 0.135 = 3.148$

$x_4 = 3.148 - \frac{(3.148-3)^3-\sin(3.148)}{3(3.148-3)^2-\cos(3.148)} = 3.148 - \frac{0.0036}{1.06} = 3.148 - 0.0034 = 3.138$

$x_5 = 3.138 - \frac{(3.138-3)^3-\sin(3.138)}{3(3.138-3)^2-\cos(3.138)} = 3.138 - \frac{0.000056}{1.057} = 3.138 - 0.00004 = 3.138$

8. Incorrect

The solution of $(x-6)^3 = \ln(x)$ is: x=7.256

$f(x) = (x-6)^3 - \ln(x)$

$f'(x) = 3(x-6)^2 - \frac{1}{x}$

$x_1 = x_0 - \frac{f(x_0)}{f'(x_0)}$

We decide the initial value as 10.

$x_1 = 10 - \frac{(10-6)^3-\ln(10)}{3(10-6)^2-\frac{1}{10}} = 10 - \frac{61.69}{47.9} = 10 - 1.28 = 8.711$

$x_2 = 8.711 - \frac{(8.711-6)^3-\ln(8.711)}{3(8.711-6)^2-\frac{1}{8.711}} = 8.711 - \frac{17.78}{21.94} = 8.711 - 0.81 = 7.9$

$x_3 = 7.9 - \frac{(7.9-6)^3-\ln(7.9)}{3(7.9-6)^2-\frac{1}{7.9}} = 7.9 - \frac{4.81}{10.72} = 7.9 - 0.448 = 7.45$

$x_4 = 7.45 - \frac{(7.45-6)^3-\ln(7.45)}{3(7.45-6)^2-\frac{1}{7.45}} = 7.45 - \frac{1.06}{6.2} = 7.45 - 0.17 = 7.28$

$x_5 = 7.28 - \frac{(7.28-6)^3-\ln(7.28)}{3(7.28-6)^2-\frac{1}{7.28}} = 7.28 - \frac{0.12}{4.79} = 7.28 - 0.025 = 7.256$

9. Correct

The solution of $2(x-9)^3 = \ln(x) + 2x$ is: x=11.33

$f(x) = 2(x-9)^3 - \ln(x) - 2x$

$f'(x) = 6(x-6)^2 - \frac{1}{x} - 2$

$x_1 = x_0 - \frac{f(x_0)}{f'(x_0)}$

We decide the initial value as 12.

$x_1 = 12 - \frac{2(12-9)^3 - \ln(12) - 2(12)}{6(12-6)^2 - \frac{1}{12} - 2} = 12 - \frac{12}{213.91} = 12 - 0.12 = 11.871$

$x_2 = 11.871 - \frac{2(11.871-9)^3 - \ln(11.871) - 2(11.871)}{6(11.871-6)^2 - \frac{1}{11.871} - 2} = 11.871 - \frac{21.13}{204.754} = 11.871 - 0.10 = 11.768$

$x_3 = 11.768 - \frac{2(11.768-9)^3 - \ln(11.768) - 2(11.768)}{6(11.768-6)^2 - \frac{1}{11.768} - 2} = 11.768 - \frac{16.42}{197.54} = 11.768 - 0.083 = 11.685$

$x_4 = 11.685 - \frac{2(11.685-9)^3 - \ln(11.685) - 2(11.685)}{6(11.685-6)^2 - \frac{1}{11.685} - 2} = 11.685 - \frac{12.887}{191.83} = 11.685 - 0.067 = 11.617$

$x_5 = 11.617 - \frac{2(11.617-9)^3 - \ln(11.617) - 2(11.617)}{6(11.617-6)^2 - \frac{1}{11.617} - 2} = 11.617 - \frac{10.19}{187.27} = 11.617 - 0.054 = 11.56$

$x_6 = 11.56 - \frac{2(11.56-9)^3 - \ln(11.56) - 2(11.56)}{6(11.56-6)^2 - \frac{1}{11.56} - 2} = 11.56 - \frac{8.11}{183.62} = 11.56 - 0.044 = 11.519$

$x_7 = 11.519 - \frac{2(11.519-9)^3 - \ln(11.519) - 2(11.519)}{6(11.519-6)^2 - \frac{1}{11.519} - 2} = 11.519 - \frac{6.49}{180.68} = 11.519 - 0.035 = 11.48$

$x_8 = 11.48 - \frac{2(11.48-9)^3 - \ln(11.48) - 2(11.48)}{6(11.48-6)^2 - \frac{1}{11.48} - 2} = 11.48 - \frac{5.22}{178.31} = 11.48 - 0.029 = 11.45$

$x_9 = 11.45 - \frac{2(11.45-9)^3 - \ln(11.45) - 2(11.45)}{6(11.45-6)^2 - \frac{1}{11.45} - 2} = 11.45 - \frac{4.21}{176.39} = 11.45 - 0.023 = 11.43$

$x_{10} = 11.43 - \frac{2(11.43-9)^3 - \ln(11.43) - 2(11.43)}{6(11.43-6)^2 - \frac{1}{11.43} - 2} = 11.43 - \frac{3.4}{174.83} = 11.43 - 0.019 = 11.41$

$x_{11} = 11.41 - \frac{2(11.41-9)^3 - \ln(11.41) - 2(11.41)}{6(11.41-6)^2 - \frac{1}{11.41} - 2} = 11.41 - \frac{2.76}{173.56} = 11.41 - 0.015 = 11.39$

....

....

$x_{22} = 11.332 - \frac{2(11.332-9)^3 - \ln(11.332) - 2(11.332)}{6(11.332-6)^2 - \frac{1}{11.33} - 2} = 11.332 - \frac{0.29}{168.54} = 11.332 - 0.0017 = 11.331$

Chapter 3. I. d. Problems in contextual situations, including related rates and optimization problems

For first 4 questions we assume that all variables depend on z.

1. Correct

If $x^3 - 2x + y = 1$ the first derivative with respect to z is: $3x^2x' - 2x' + y' = 0$
If we differentiate $x^3 - 2x + y = 1$ with respect to z, we have: $3x^2x' - 2x' + y' = 0$

2. Incorrect

If $2x^5 + 3x^4 - 4x + y^2 = 10$ the first derivative with respect to z is: $10x^4x' + 12x^3x' - 4x' + 2yy' = 0$

If we differentiate $2x^5 + 3x^4 - 4x + y^2 = 10$ with respect to z, we have:

$10x^4x' + 12x^3x' - 4x' + 2yy' = 0$

3. Correct

If $x^3 + 3y^2 - x^2 = \sin(x)$ the first derivative with respect to z is:

$3x^2x' + 6yy' - 2xx' = x'\cos(x)$

If we differentiate $x^3 + 3y^2 - x^2 = \sin(x)$ with respect to z, we have:

$3x^2x' + 6yy' - 2xx' = x'\cos(x)$

4. Incorrect

If A=πR^2 the first derivative with respect to z is: $A' = 2\pi RR'$

If we differentiate A=πR^2 with respect to z, we have:

$A' = 2\pi RR'$

5. Correct

We have a rectangular yard with perimeter of 300 m. It has to be fenced. The dimensions of the yard that will give us the greatest area are: x=y=75m.

Let's suppose the length equals x and width equals y.

Then we have:

$Area(A) = xy, and\ Perimeter(P) = 2x + 2y = 300$

$2x = 300 - 2y$

$x = \frac{300-2y}{2} = 150 - y$

We substitute x in formula for area.

$A = xy = (150 - y)y = 150y - y^2$

To find the maximum value of the Area we need to find the y values for which the derivative of function A(y) is zero.

$A(y) = 150y - y^2$

$A' = (150y)' - (y^2)' = 150 - 2y = 0$

$150 = 2y$

$So, y = \frac{150}{2} = 75m$

$x = 150 - y = 150 - 75 = 75m$

6. Incorrect

We have to build a box with the length of the base four times the width. The height is the 3 times length minus 10 cm. The dimensions that will minimize the volume are: Length= 7.2 cm , Width= 1.8 cm , height= 11.6 cm

Length=x, Width=y, height=h

$Area(A) = xy, where\ x = 4y$

$Area(A) = xy = (4y)y = 4y^2$

$h = 3x - 10 = 12y - 10$

$Volume(V) = A * h = (4y^2)h = 4y^2(3x - 10) = 2y^2(12y - 10) = 24y^3 - 20y^2$

$V' = (24y^3)' - (20y^2)' = 72y^2 - 40y = 0$

$72y^2 - 40y = 8y(9y - 5) = 0$

$y = 0$ not real

$y = Width = \frac{9}{5}\ cm = 1.8\ cm$ then:

$x = Length = 4(1.8) = 7.2\ cm\ and, h = Height = 3(7.2) - 10 = 11.6\ cm$

7. Correct

We have to build a box with a base that have length two times the width, and we have 25 m square of material. The dimensions for the maximum volume are:

Length= 2.88 m , Width= 1.44 m and Height= 1.93 m

We have:

Length = x ; Width = y ; Height = h

$x = 2y$

$Surface\ area(SA) = 2xy + 2xh + 2yh = 25\ cm^2$

We isolate h

$2xy + 2xh + 2yh = 25$

$2xh + 2yh = 25 - 2xy = 25 - 2(2y)y = 25 - 4y^2$

$2xh + 2yh = 25 - 4y^2$

$2(2y)h + 2yh = 25 - 4y^2$

$4yh + 2yh = 25 - 4y^2$

$6yh = 25 - 4y^2$

$h = \frac{25-4y^2}{6y}$

So, the volume can be written as:

$V = xyh = (2y)y\left(\frac{25-4y^2}{6y}\right) = y(\frac{25-4y^2}{3})$

$V(y) = y(\frac{25-4y^2}{3})$

$V'(y) = y'\left(\frac{25-4y^2}{3}\right) + y(\frac{25-4y^2}{3})' = \left(\frac{25-4y^2}{3}\right) + \frac{y(-8y)}{3} = \frac{25-4y^2-8y^2}{3} = \frac{25-12y^2}{3}$

So:

$V'(y) = \frac{25}{3} - 4y^2 = 0$

$\frac{25}{3} = 4y^2$

$y^2 = \frac{25}{12}; then\ y = \mp\sqrt{\frac{25}{12}} = \mp 1.44$

y can't be negative so, y=1.44m

$x = 2y = 2 * 1.44 = 2.88\ m$

$h = \frac{25-4y^2}{6y} = \frac{25-4(1.44)^2}{6(1.44)} = \frac{25-8.29}{8.64} = \frac{16.71}{8.64} = 1.93\ m$

8. Incorrect

We have to build a tunnel that has a cylinder shape that has 50 Liters in volume. The dimensions of the tunnel in order to have the smallest surface area are: R= 29.24 cm And h = 18.62 cm.

$V = \pi R^2 h = 50{,}000\ cm^3\ so, h = \frac{50{,}000}{\pi R^2}$

$SA = 2\pi R^2 + 2\pi Rh = 2\pi R^2 + \frac{2\pi R(50{,}000)}{\pi R^2} = 2\pi R^2 + \frac{100{,}000}{R}$

$SA = 2\pi R^2 + \frac{100{,}000}{R} = SA(R)$

So:

$(SA)' = (2\pi R^2 + \frac{100{,}000}{R})' = 2\pi(2R) + (100{,}000R^{-1})' = 4\pi R - 100{,}000R^{-2} = 0$

$4\pi R - 100{,}000R^{-2} = 0$

$4\pi R = \frac{100{,}000}{R^2}$

$4\pi R^3 = \frac{100{,}000}{} so, R^3 = \frac{100{,}000}{4\pi} = 7961.78$

$R = \sqrt[3]{7961.78} = 19.96\ cm$

$h = \frac{50{,}000}{\pi R^2} = \frac{50{,}000}{\pi(19.96)^2} = \frac{50{,}000}{1250.98} = 39.96\ cm$

9. Correct

The profit relation for a company is $P = Revenue - expenses = \frac{1230}{p} - \frac{550}{(p)^2}$ The price per unit (p) for maximum profit P is: $p = \$0.89$

The first derivative of Profit is:

$P'(p) = (\frac{1230}{p})' - \left(\frac{550}{p^2}\right)' = 1230(p^{-1})' - 550(p^{-2})' = -1230p^{-2} + 1100p^{-3}$

$P'(p) = -1230p^{-2} + 1100p^{-3} = 0$

$1230p^{-2} = 1100p^{-3}$

$1230 = \frac{1100}{p}$

$p = \frac{1100}{1230} = \$0.89$

10. Incorrect

We need to fence two adjacent lots of land. We have 240 m of fence. The dimensions x and y to maximize the area are: x = 3 m and y = 4 m.

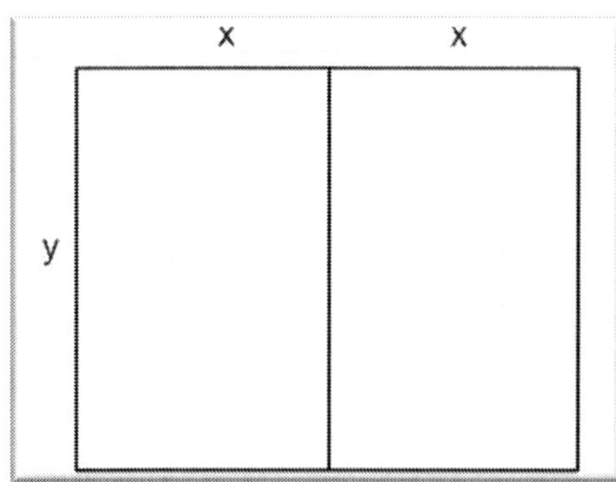

$P = 3y + 4x = 240\ m$

$4x = 240 - 3y$

$x = \frac{240-3y}{4} = 60 - 0.75y$

$Area\ (A) = 2xy = 2(60 - 0.75y)y = 120y - 1.5y^2 = A(y)$

$A'(y) = (120y)' - (1.5y^2)' = 120 - 3y$

$A'(y) = 0 = 120 - 3y$

$120 = 3y\ so,$

$y = \frac{120}{3} = 40\ m$

$x = 60 - 0.75y = 60 - 0.75(40) = 60 - 30 = 30\ m$

CHAPTER 4

Integration

"If I have seen further than others,
it is by standing upon the shoulders of giants."

Isaac Newton (1642-1727)

(conceived the ideas of differential and integral calculus)

Questions to be answered:

- Sagrada Familia is one of the most interesting churches in the world. It is built in Barcelona, Spain. The name of the architect that designed it is…….
- This peak is situated in Grater Vancouver area. It has an elevation of 1,200 m. It is a great touristic attraction all year long. The name of the mountain is……… Mountain
- This animal can freeze without dying. What animal is this?
- 37 stars were programed on the Apollo spacecraft's computer. Three were not real stars. One was star number 20 after Edward White II written backwards. What was the name of the star?
- This was Rome's closest port, situated at the shores of Mediterranean Sea. From here the cargo was sent up to Rome by barge, following the Tiber river. It was also a military port. The name of this port was…
- Fallingwater, a "weekend home" for Edgar Kaufmann's family, it is one of the most famous buildings in the world. It was built in 1935-1936. The architect was the famous Frank Lloyd…
- This mountain is the highest in Switzerland at 4364 m tall. It is situated near the Swiss – Italian border. Its name is Monte …
- On July 20th 1969, Neil Armstrong and Buzz Aldrin were the first humans that landed on the surface of the Moon. Armstrong noted that the soil is "almost like a ……."
- Only the members of the Roman senate were allowed to wear a tunic called a 'toga,' with a broad -purple stripe. This purple stripe was called *latus…*

4. A. Definition of an integral and notation

Theory and Examples

Simplistically, an integral is a weighted sum of the values of the function multiplied with the infinitesimal widths dx.

There are two types of integrals; indefinite and definite.

The symbol for indefinite integral is $\int$. The symbol for the definite integral is $\int_a^b$, where a and b are the ends of the interval for which the integral is calculated.

The mathematical process that helps us find an integral it is called integration.

If the variable of the function is x, the indefinite integral is symbolized $\int f(x)dx$.

The function f(x) under the integral sign is called integrand, and x is called the integration variable

Please note that dx is an infinitesimal quantity. To integrate something means as well that we add many products formed by the value of the function multiplied with the infinitesimal quantity dx.

But what do these pairs of f(x) times dx represent? If we try to visualize them on a system of axes, they represent the very small area of a rectangle with one side being dx and the height being f(x).

EXAMPLE

In the graph showed below, we want to calculate the area of the trapezoid ALQK. We split the trapezoid into rectangles. In this case, the number is only 3 for clarity. The rectangles have the base dx=1 unit. The top side of the rectangles go through the value of f(x) where x is the middle of the base dx. The heights of the rectangles

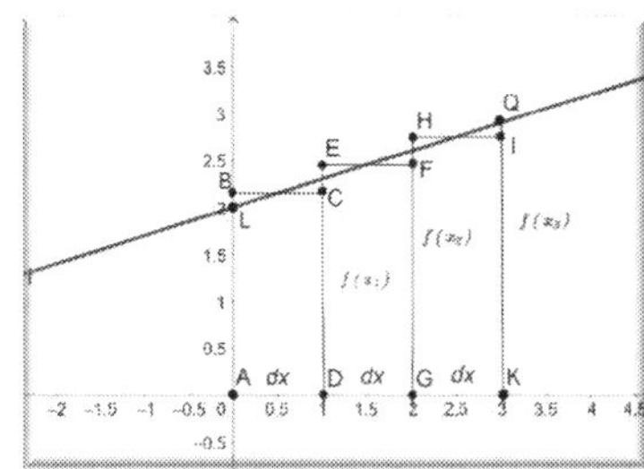

$DC = f(x_1); GF = f(x_2); KI = f(x_3)$

The area of ALQK, is approximately the sum of the areas of ABCD, DEFG, GHIK.

We write the sum as:

$Area = \sum_{i=1}^{3} f(x_i)dx = f(x_1)dx + f(x_2)dx + f(x_3)dx$

As we said before, dx is very small. So, to find the best approximation for area, we have to split the trapezoid into very small rectangles. We have then to add a lot of very small areas. We can approximate the area with:

$Area \cong \sum_{i=1}^{n} f(x_i)dx = f(x_1)dx + f(x_2)dx + \cdots f(x_n)dx$ where n is a very big number but finite.

When dx becomes infinitesimal, n approaches infinity so, instead of adding multiple small areas we will use integration instead. $Area = \lim_{n\to\infty} \sum_{i=1}^{n} f(x_i)dx = \int_0^n f(x_i)dx$

In the next chapters we will discuss in more detail what this equation means, and how we should understand the many, many super-small intervals dx.

Chapter 4. A. Definition of an integral and notation

Sagrada Familia is one of the most interesting churches in the world. It is built in Barcelona, Spain. The name of the architect that designed it is ....

Determine which answer is correct. In the table at the bottom of the page cross off all the letters of the correct answers. The word that remains is the answer.

1) An integral is a weighted sum of the values of the function times the infinitesimal widths dx.

2) The number of the infinitesimal widths equals 100.

3) The notation of definite integral in the interval [a,b] is: $\int_a^b f(x)dx$.

4) The sign $\int$ represents the product between $f(x)$ and dx.

5) The notation of indefinite integral is: $\int f(x^2)dx$.

6) The function $f(x)$ under the integral sign it is called quotient.

7) The first documented technique that tried to calculate the integral was used by the ancient Greek astronomer Euxodus around 370 BC.

8) Other method used is called Monte Carlo method.

1	2	3	4	5	6	7	8
E	G	R	A	U	D	U	I

4.B. Definite and indefinite integrals

Theory and Examples

If we have a function $f(x)$, an anti-derivative of this function is any function $F(x)$ such that $F'(x) = f(x)$

The indefinite integral of $f(x)$ is:

$\int f(x)dx = F(x) + C$ where, $F(x)$ is the anti-derivative of $f(x)$, and C is any constant.

EXAMPLE

If $f(x) = 2x$, $\int f(x)dx = x^2 + C = F(x) + C$,

$F'(x) = (x^2)' = 2x = f(x)$

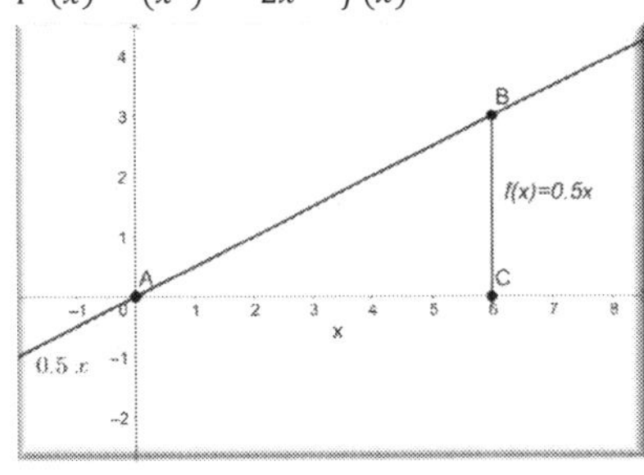

Finding Area Problem

We want to calculate the area of the triangle ABC. The function that constitutes the hypotenuse AB is:

$f(x) = \frac{1}{2}x$

The base of the triangle AC is x.

The height CB is $f(x) = 0.5x$

$Area = \frac{AC*BC}{2} = \frac{1}{2}\left(\frac{1}{2}x\right)x = \frac{1}{4}x^2$

What is the derivative of $\frac{1}{4}x^2$?

$(\frac{1}{4}x^2)' = \frac{1}{4} * 2x = \frac{1}{2}x = f(x)$, so, $\frac{1}{4}x^2 = F(x)$

Area of the surface that is below y=f(x) in between zero and 6 is the definite integral:

$\int_0^6 f(x)dx = F(x)$ between x=0 and x=6

In the future, we will discuss in more detail how to calculate the indefinite and definite integrals.

Some indefinite integrals:

$\int x^3 dx = \frac{x^4}{4} + C$

$\int \frac{dx}{x} = \ln|x| + C$

$\int e^x dx = e^x + C$

$\int \ln(x)\, dx = xln(x) - x + C$

$\int \sin(x)\, dx = -\cos(x) + C$

$\int a^x dx = \frac{a^x}{\ln(a)} + C$

$\int \cos(x)\, dx = sin(x) + C$

$\sqrt{x} = x^{\frac{1}{2}}\ so,$

$\int \sqrt{x} dx = \int x^{\frac{1}{2}} dx = \frac{2x^{\frac{3}{2}}}{3} + C$

Chapter 4. B. Definite and indefinite integrals

This peak is situated in the Greater Vancouver area. It has an elevation of 1,200 m. It is a great touristic attraction all year long. The name of the mountain is……… Mountain

Determine which answer is correct. In the table at the bottom of the page cross off all the letters for the correct answers. The word that remains is the answer.

1) If $f(x) = 3x^2 - e^x$, then $\int f(x)dx = x^3 - e^x + C$

2) If $f(x) = 2x^4 + \ln(x)$, then $\int f(x)dx = 8x^3 - e^x + C$

3) If $f(x) = \ln(x) - 2^x$, then $\int f(x)dx = \frac{1}{x} + C$

4) If $f(x) = 1 + \ln(x)$, then $\int f(x)dx = xln(x) + C$

5) If $f(x) = 2 - cos\,(x)$, then $\int f(x)dx = 2x - ln(x) + C$

6) If $f(x) = x^2 - 2x$, then $\int_0^x f(x)dx = \frac{1}{3}x^3 - x^2 + C$

7) If $f(x) = 2x^2 + 3x - 4$, then $\int_0^x f(x)dx = 6$

8) If $f(x) = x^2 - 4\cos(x)$, then $\int_0^6 f(x)dx = -7\frac{1}{3}$

9) If $f(x) = \ln(x) + 3x$, then $\int_0^4 f(x)dx = 24.069$

10) If $f(x) = 2x + \sqrt{x}$, then $\int_0^3 f(x)dx = 5$

1	2	3	4	5	6	7	8	9	10
A	G	R	E	O	N	U	S	I	E

4.C. Approximations-Riemann sum, rectangle method, trapezoidal method

Theory and Examples

Riemann sum helps us approximate the definite integral $\int_a^b f(x)dx$ with the area beneath the graph of a function and for x values in a closed interval [a,b].

EXAMPLE

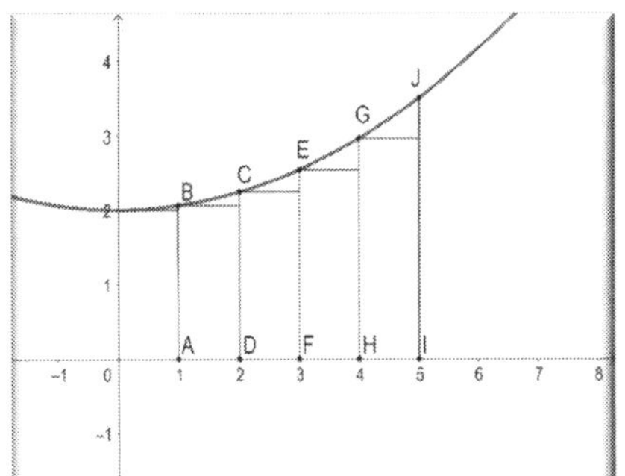

Let's have a=0 and b=5. The function is $f(x) = 0.06x^2 + 2$. We create rectangles that touch the graph at the top left corner. We add the areas of these rectangles. This is called Riemann left sum. Coordinates of these top left corner points are shown below.

Point	X	F(x)
2	0	2
B	1	2.06
C	2	2.24
E	3	2.54
G	4	2.96

As it can be seen in the table, the distance between each consecutive x value is $\Delta x = 1$

The sum of these areas is:

$\int_0^4 f(x)dx \cong \sum_{i=0}^4 f(x_i) * \Delta x_i = f(0) * 1 + f(1) * 1 + f(2) * 1 + f(3) * 1 + f(4) * 1 = 2 + 2.06 + 2.24 + 2.54 + 2.96 = 11.8$

If we want a better approximation of the area beneath the graph, we need to choose much thinner rectangles with Δx much smaller.

Trapezoidal method helps us find the integral $\int_a^b f(x)dx$ by approximating it with the area beneath the graph of a function by partitioning the area in small trapezoids and calculate the total area of these trapezoids.

EXAMPLE

Let's use a=0 and b=4. We are using the same function as before: $f(x) = 0.06x^2 + 2$.

The formula for the area of a trapezoid is:

$Area = \frac{(Big\ base + small\ base) * height}{2}$

In this case height Δx is 1.

We approximate the integral with the sum of all the trapezoids the area is portioned in.

We consider four trapezoids 1 to 4.

$\int_1^4 f(x)dx \cong \sum_1^4 \frac{(Big\ base_i + small\ base_i) * \Delta x_i}{2}$

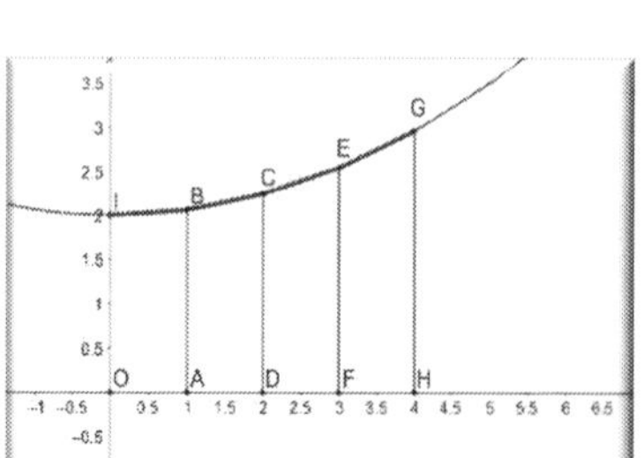

$$Area_{IBAO} = Area_1 = \frac{(OI+AB)*AO}{2} = \frac{(2+2.06)*1}{2} = 2.03$$

$$Area_{BADC} = Area_2 = \frac{(AB+DC)*AD}{2} = \frac{(2.06+2.24)*1}{2} = 2.15$$

$$Area_{DCEF} = Area_3 = \frac{(DC+EF)*DF}{2} = \frac{(2.24+2.54)*1}{2} = 2.39$$

$$Area_{DCEF} = Area_4 = \frac{(FEGH)*FH}{2} = \frac{(2.54+2.96)*1}{2} = 2.75$$

$\int_1^4 f(x)dx \cong \sum_1^4 \frac{(Big\ base_i+small\ base_i)*\Delta x_i}{2} = \frac{(OI+AB)*AO}{2} + \frac{(AB+DC)*AD}{2} + \frac{(DC+EF)*DF}{2} + \frac{(FEGH)*FH}{2} = 2.03 + 2.15 + 2.39 + 2.75 = 9.32$

EXAMPLE

If $f(x) = 3x^3 - 2x^2 + 1$, using trapezoidal method, find the integral $\int_{-2}^{2} f(x)dx$

The interval $\Delta x = 0.5$

$\int_{-2}^{2} f(x)dx \cong \sum_{i=-2}^{i=2} \frac{f(x_i)+f(x_{i+1})}{2}\Delta x_i = \frac{f(-2)+f(-1.5)}{2} * 0.5 + \frac{f(-1.5)+f(-1)}{2} * 0.5 + \frac{f(-1)+f(-0.5)}{2} * 0.5 + \frac{f(-0.5)+f(0)}{2} * 0.5 + \frac{f(0)+f(0.5)}{2} * 0.5 + \frac{f(0.5)+f(1)}{2} * 0.5 + \frac{f(1)+f(1.5)}{2} * 0.5 + \frac{f(1.5)+f(2)}{2} * 0.5$

$f(-2) = 3(-2)^3 - 2(-2)^2 + 1 = -24 - 8 + 1 = -31$

$f(-1.5) = 3(-1.5)^3 - 2(-1.5)^2 + 1 = 3(-3.375) - 2(2.25) + 1 = -10.125 - 4.5 + 1 = -13.625$

$f(-1) = 3(-1)^3 - 2(-1)^2 + 1 = -3 - 2 + 1 = -4$

$f(-0.5) = 3(-0.5)^3 - 2(-0.5)^2 + 1 = 3(-0.125) - 2(0.25) + 1 = -0.375 - 0.5 + 1 = 0.13$

$f(0) = 3(0)^3 - 2(0)^2 + 1 = 1$

$f(0.5) = 3(0.5)^3 - 2(0.5)^2 + 1 = 3(0.125) - 2(0.25) + 1 = 0.375 - 0.5 + 1 = 0.875$

$f(1) = 3(1)^3 - 2(1)^2 + 1 = 3 - 2 + 1 = 2$

$f(1.5) = 3(1.5)^3 - 2(1.5)^2 + 1 = 3(3.375) - 2(2.25) + 1 = 10.125 - 4.5 + 1 = 6.625$

$f(2) = 3(2)^3 - 2(2)^2 + 1 = 24 - 8 + 1 = 17$

$\int_{-2}^{2} f(x)dx \cong \sum_{i=-2}^{i=2} \frac{f(x_i)+f(x_{i+1})}{2}\Delta x_i = \frac{f(-2)+f(-1.5)}{2} * 0.5 + \frac{f(-1.5)+f(-1)}{2} * 0.5 + \frac{f(-1)+f(-0.5)}{2} * 0.5 + \frac{f(-0.5)+f(0)}{2} * 0.5 + \frac{f(0)+f(0.5)}{2} * 0.5 + \frac{f(0.5)+f(1)}{2} * 0.5 + \frac{f(1)+f(1.5)}{2} * 0.5 + \frac{f(1.5)+f(2)}{2} * 0.5 = \frac{-27-13.625}{2} * 0.5 + \frac{-13.625-4}{2} * 0.5 + \frac{-4-0.125}{2} * 0.5 + \frac{-0.125+1}{2} * 0.5 + \frac{1+0.875}{2} * 0.5 + \frac{0.875+2}{2} * 0.5 + \frac{2+6.625}{2} * 0.5 + \frac{6.625+17}{2} * 0.5 = -11.16 - 4.40 - 0.969 + 0.218 + 0.468 + 0.718 + 2.156 + 5.9 = -7$

Chapter 4. C. Approximations-Riemann sum, rectangle method, trapezoidal method

This animal can freeze without dying. What is the name of this animal?

Determine which answer is correct. In the table at the bottom of the page cross off all the letters for the correct answers. The word that remains is the answer.

1) If $f(x) = 2x^2 - 3x$, using Riemann left sum, the integral $\int_{-2}^{3} f(x)dx = 20$
The interval $\Delta x = 1$

2) If $f(x) = 4x^3 + 3x^2 - 2x + 1$, using Riemann left sum, the integral $\int_{-3}^{4} f(x)dx \cong 20$
The interval $\Delta x = 1$

3) If $f(x) = 3x^3 - 2x^2 + x - 1$, using Rectangle Method, the integral $\int_{-2}^{3} f(x)dx \cong 21.875$
The interval $\Delta x = 1$

4) If $f(x) = x^2 + 2x + \sqrt[3]{x}$, using Rectangle Method, the integral $\int_{-3}^{3} f(x)dx \cong 2.82$
The interval $\Delta x = 1$

5) If $f(x) = 4x^3 + \sin(x)$, using Riemann left sum, the integral $\int_{-3}^{3} f(x)dx \cong -108.05$
The interval $\Delta x = 1$

6) If $f(x) = 3x^3 - 2x^2 + 1$, using left side Riemann sum, the integral $\int_{-2}^{2} f(x)dx \cong 3$
The interval $\Delta x = 1$

7) If $f(x) = 3x^3 - 2x^2 + 1$, using left side Riemann sum, the integral $\int_{-2}^{2} f(x)dx \cong -17.12$
The interval $\Delta x = 0.5$

8) If $f(x) = 3x^3 - 2x^2 + 1$, using trapezoidal method, the integral $\int_{-2}^{2} f(x)dx \cong 66$
The interval $\Delta x = 1$

9) If $f(x) = 3x^3 - 2x^2 + 1$, using trapezoidal method, the integral $\int_{-2}^{2} f(x)dx \cong -6.83$
The interval $\Delta x = 0.5$

10) If $f(x) = x^2 - x - 6$, using trapezoidal method, the integral $\int_{-2}^{2} f(x)dx \cong 5.48$
The interval $\Delta x = 0.5$

1	2	3	4	5	6	7	8	9	10
A	F	E	R	K	O	L	G	U	S

4.D. Fundamental Theorem of Calculus

Theory and Examples

The fundamental theorem of calculus says that:

$$\int_a^b f(x)dx = \int_0^b f(x)dx - \int_0^a f(x)dx = F(b) - F(a), where\ F'(x) = f(x)$$

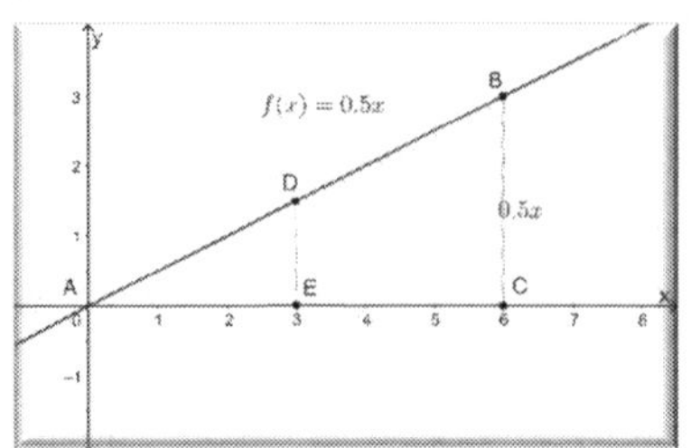

Before we explain the formula, we go back to the area problem in 4.B.

Area Problem extended

We want to calculate the area of EDBC. First, we calculate the area of the triangle ABC. The function that constitutes the hypotenuse AB is:

$f(x) = \frac{1}{2}x$

The base of the triangle AC is x.
The height CB is $f(x) = 0.5x$

$Area = \frac{1}{2}AC * CB = \left(\frac{1}{2}x\right)x = \frac{1}{4}x^2$

What is the derivative of $\frac{1}{4}x^2$?

$(\frac{1}{4}x^2)' = \frac{1}{4} * 2x = \frac{1}{2}x = f(x)$, so,

$\frac{1}{4}x^2 = F(x) = f'(x)$

Area of the surface ABC that is below y=f(x) in between zero and 6 is the definite integral:

$\int_0^6 f(x)dx = F(x)$ between x=0 and x=6

$F(6) = \frac{1}{4}(6)^2 = \frac{36}{4} = 9$

$F(0) = 0$

Second, we calculate the area of the triangle ADE.
The function that constitutes the hypotenuse AD is:

$f(x) = \frac{1}{2}x$

Area of the surface ADE that is below y=f(x) in between zero and 3 is the definite integral:

$\int_0^3 f(x)dx = F(x)$ between x=0 and x=3

$F(3) = \frac{1}{4}(3)^2 = \frac{9}{4} = 2.25$

$Geometric\ Area_{EDBC} = \frac{(Big\ base+small\ base)*Height}{2} = \frac{(3+1.5)*3}{2} = 6.75$

What we notice is that if we subtract $F(6) - F(3) = 9 - 2.25 = 6.75$ equal with the area of EDBC

So, the area situated below the function $f(x) = \frac{1}{2}x$ between x=3 and x=6 is given by:

$\int_3^6 f(x)dx = \int_0^6 f(x)dx - \int_0^3 f(x)dx = F(6) - F(3) = 9 - 2.25 = 6.75, where\ F'(x) = f(x)$

EXAMPLE

Find the area under the graph of the function $f(x) = 0.3x^2 + 1$ between x=-1 and x=3 using the Fundamental Theorem of Calculus.

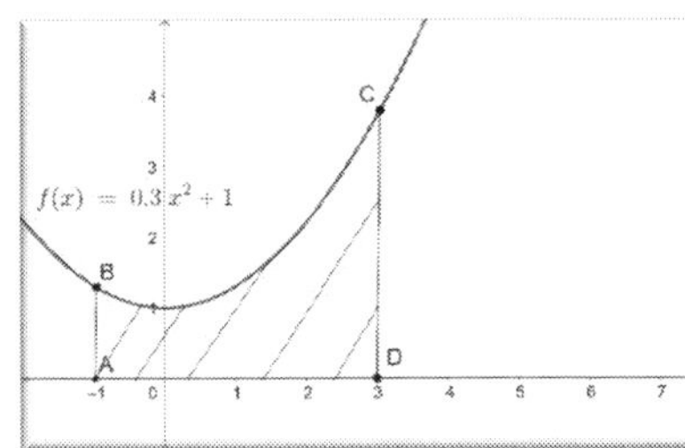

$\int_{-1}^{3} f(x)dx = F(3) - F(-1)$

$F(x) = \int f(x)dx = \frac{0.3}{3}x^3 + x + C$

So,

$F(3) = \frac{0.3}{3}(3)^3 + (3) + C = 5.7 + C$

$F(-1) = \frac{0.3}{3}(-1)^3 + (-1) + C = -1.1 + C$

So,

$\int_{-1}^{3} f(x)dx = F(3) - F(-1) = 5.7 + C - (-1.1) - C = 6.8$

EXAMPLE

Find the area below the function $f(x) = \frac{2}{x} - 2x + 5$ for x=1 and x=4

First,

We calculate the indefinite integral of the function.

$F(x) = \int f(x)dx = \int \frac{2}{x}dx - \int 2xdx + \int 5dx = 2\ln(x) - \frac{2}{2}x^2 + 5x + C = 2\ln(x) - x^2 + 5x$

Second,

We calculate the values of the F(x) for x=4 and x=1

$F(4) = 2\ln(4) - (4)^2 + 5(4) = 2\ln(4) - 16 + 20 = 2\ln(4) + 4 = 2(1.38) + 4 = 2.77 + 4 = 6.77$

$F(1) = 2\ln(1) - (1)^2 + 5(1) = -1 + 5 = 4$

Third,

We apply the Fundamental Theorem of Calculus:

$\int_1^4 f(x)dx = F(4) - F(1) = 6.77 - 4 = 2.77$

Chapter 4. D. Fundamental Theorem of Calculus

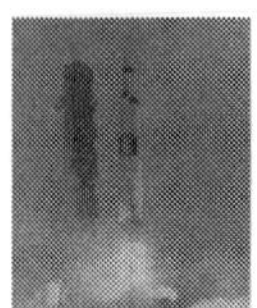

37 stars were programed on the Apollo spacecraft's computer. Three were not real stars. One was star number 20 after Edward White II written backwards. What was the name of the star?

Determine which answer is correct. In the table at the bottom of the page cross off all the letters of the correct answers. The word that remains is the answer.

1) If $f(x) = 3x^2 + 3x - 2$ then, the $\int_{-4}^{5} f(x)dx = 184.5$

2) If $f(x) = x^2 + 3\sqrt{x}$ then, the $\int_{1}^{5} f(x)dx = 45$

3) If $f(x) = 4x - 3$ then, the $\int_{-4}^{5} f(x)dx = 18$

4) If $f(x) = 5x + 6\sqrt{x}$ then, the $\int_{2}^{7} f(x)dx = 175.3 + C$

5) If $f(x) = \frac{7}{x} + x - 3$ then, the $\int_{1}^{5} f(x)dx = 5$

6) If $f(x) = \frac{4+x^2}{x} - 3x + 7$ then, the $\int_{2}^{8} f(x)dx = 7$

7) If $f(x) = 4x + e^x$ then, the $\int_{-2}^{6} f(x)dx = 467.28$

8) If $f(x) = e^x + \sin(x)$ then, the $\int_{1}^{6} f(x)dx = 7.28$

9) If $f(x) = 2e^x + 3x^2 - 4x + 5$ then, the $\int_{-3}^{8} f(x)dx = 28$

10) If $f(x) = x^2 - 3x + 7$ then, the $\int_{-3}^{3} f(x)dx = 60$

1	2	3	4	5	6	7	8	9	10
A	D	N	L	O	C	A	E	S	R

4.E. Integration

a. Antiderivatives of functions

Theory and Examples

The antiderivative F(x) is reverse of the derivative. We know that $F'(x) = original\ function\ f(x)$ Suppose that $f(x) = 3x + 1$. Then the derivate of the antiderivative $F'(x) = 3x + 1 = f(x)$ will be equal with the original function.

The indefinite integral of the function f(x) is $\int f(x)dx = \frac{3}{2}x^2 + x + C = F(x)$

We can see that, if we derivate F(x) we get f(x).

$F'(x) = (\frac{3}{2}x^2 + x + C)' = \frac{3*2}{2}x + 1 + 0 = 3x + 1 = f(x)$

The antiderivative is connected with the definite integral through the Fundamental Theorem of Calculus.

If we have an interval [a,b].

$\int_a^b f(x)dx = F(b) - F(a)\ where\ F(b)\ and\ F(a)$ are the antiderivatives of f(x) at x=b and x=b.

EXAMPLE

If the acceleration of a car is given by the function $f(t) = 5t + 3, where\ t\ is\ time,$

the velocity is given by the antiderivative $F(t) = \frac{5}{2}t^2 + 3t = \int f(t)dt = V(t)$

The derivative of F(t) is:

$F(t)' = V'(t) = (\frac{5}{2}t^2 + 3t)' = \frac{5*2}{2}t + 3 = 5t + 3 = f(t) = acceleration$

The distance is given by:

$s(t) = \int V(t)dt = \frac{5}{2*3}t^3 + \frac{3}{2}t^2 + C = \frac{5}{6}t^3 + \frac{3}{2}t^2 + C$

$s'(t) = (\frac{5}{6}t^3 + \frac{3}{2}t^2 + C)' = \frac{5*3}{6}t^2 + \frac{3*2}{2}t + 0 = \frac{5}{2}t^2 + 3t = V(t)$

Chapter 4. E. a. Antiderivatives of functions

This was Rome's closest port situated at the shores of the Mediterranean Sea. From here, cargo was sent up to Rome by barge, following the Tiber river. It was also a military port. The name of this port was...

Determine which answer is correct. In the table at the bottom of the page cross off all the letters of the correct answers. The word that remains is the answer.

1) If $f(x) = \sin(x) + 2x - \frac{3x^2+3x}{x}$
then, the antiderivative of $f(x)$ will be: $F(x) = -\cos(x) - \frac{x^2}{2} - 3x + C$

2) If $f(x) = 5x + \frac{4x^3+3x^2-x}{3}$ then,
the antiderivative of $f(x)$ will be: $F(x) = 5x - \frac{7x^2}{3} + C$

3) If $f(x) = 7\tan(x) - sec^2(x)$ then,
the antiderivative of $f(x)$ will be: $F(x) = -7\ln|\cos(x)| - \tan(x) + C$

4) If $f(x) = e^x - \frac{x^2-3x-10}{x+2} + 7$ then,
the antiderivative of $f(x)$ will be: $F(x) = e^x - x + 5 + C$

5) If $f(x) = \cos(x) - \frac{2x^2+3x}{x} + 3$ then,
the antiderivative of $f(x)$ will be: $F(x) = \sin(x) - x^2 + C$

6) If $f(x) = \frac{2x^2+3x}{x^2} - 5\sec^2(x)$ then,
the antiderivative of $f(x)$ will be: $F(x) = 2x - x^2 + C$

7) If $f(x) = \frac{sin^2(x)+cos^2(x)}{x} - e^x$ then,
the antiderivative of $f(x)$ will be: $F(x) = ln|x| - e^x + C$

8) If $f(x) = \frac{sin^2(x)-cos^2(x)}{\sin(x)+\cos(x)} + 1 + \ln(x)$ then,
the antiderivative of $f(x)$ will be: $F(x) = 2ln|x| + 1 + C$

9) If $f(x) = sin^2(x) + cos^2(x) + 1 - \ln(x)$ then,
the antiderivative of $f(x)$ will be: $F(x) = 3x - xln(x) + C$

10) If $f(x) = \frac{2}{\sqrt{x}} - \sqrt{x}$ then,
the antiderivative of $f(x)$ will be: $F(x) = 2x - \frac{3x^2}{2} + C$

1	2	3	4	5	6	7	8	9	10
P	O	L	S	A	T	H	I	K	A

4.E. Integration

b. Methods of Integration - Substitution

Theory and Examples

Sometimes, we can differentiate one of the factors of a multiplication inside the integral and get part of the other term of the product.

If we have $\int 2x(x^2+1)dx$ we notice that if we derivate x^2+1, we get 2x.

We substitute x^2+1 with u:

$u' = (x^2+1)' = 2x = \frac{du}{dx}$

So,

$dx = \frac{du}{2x}$

We have:

$\int 2x(u)\frac{du}{2x} = \int udu = \frac{u^2}{2} = \frac{(x^2+1)^2}{2} + C$

So,

$\int 2x(x^2+1)dx = \frac{(x^2+1)^2}{2} + C$

EXAMPLE

If $f(x) = (4x-3)(2x^2-3x+5)^2$ then, find $F(x)$

We substitute $2x^2-3x+5$ with u:

$u = 2x^2 - 3x + 5\ so,$

Then we calculate the derivate of u:

$u' = \frac{du}{dx} = 4x - 3$

We substitute dx with du:

$dx = \frac{du}{4x-3}$

Then we calculate the integral in u:

$\int (4x-3)u^2\frac{du}{4x-3} = \int u^2du = \frac{1}{3}u^3 + C$

We substitute back u with $2x^2-3x+5$

$F(x) = \int f(x)dx = \frac{1}{3}(2x^2-3x+5)^3 + C$

Chapter 4. E. b. Methods of Integration - Substitution

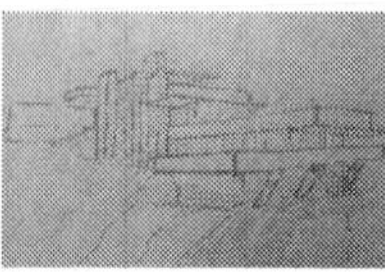

Fallingwater, a "weekend home" for Edgar Kaufmann's family, is one of the most famous buildings in the world. It was built in 1935-1936. The architect was the famous Frank Lloyd...

Determine which answer is correct. In the table at the bottom of the page cross off all the letters for the correct answers. The word that remains uncrossed is the answer.

1) If $f(x) = 4(2x+1)$ then, $F(x) = \int f(x)dx = (2x+1)^2 + C$

2) If $f(x) = 30x(5x^2-3)^2$ then, $F(x) = \int f(x)dx = (5x-3)^2 + C$

3) If $f(x) = 2xcos(x^2)$ then, $F(x) = \int f(x)dx = \sin(x^2) + C$

4) If $f(x) = \frac{5}{3x-73}$then, $F(x) = \int f(x)dx = 3x - 73 + C$

5) If $f(x) = \frac{7\ln(x)}{5x}$then, $F(x) = \int f(x)dx = \frac{7}{5} + C$

6) If $f(x) = 3x^3e^{x^4}$then, $F(x) = \int f(x)dx = \frac{3}{4}e^{x^4} + C$

7) If $f(x) = (6x-4)(3x^2-4x+5)^2$then, $F(x) = \int f(x)dx = 6x^2 - 10x + C$

8) If $f(x) = \frac{8x^3+3}{\sqrt{2x^4+3x-5}}$ then, $F(x) = \int f(x)dx = 2\sqrt{2x^4+3x-5} + C$

9) If $f(x) = \frac{9x^2-8x}{\sqrt{3x^3-4x^2+5}}$ then, $\int_1^3 f(x) = F(3) - F(1) = 4$

10) If $f(x) = \frac{1}{\sqrt{x^2+9}} = \frac{1}{\sqrt{x^2+3^2}}$ then, $F(x) = \int f(x)dx = \sqrt{2x^3+3} + C$

1	2	3	4	5	6	7	8	9	10
A	W	O	R	I	M	G	U	H	T

4. E. Methods of integration

c. by parts

Theory and Examples

When we are using the integration by parts, we are using the product rule for differentiating, and substitution rule.

Remember the product rule for differentiating a product two functions f(x) and g(x) is given by the formula:

$[f(x) * g(x)]' = f'(x) * g(x) + f(x) * g'(x)$

If we integrate this formula we get:

$\int [f(x) * g(x)]'dx = \int f'(x) * g(x)dx + \int f(x) * g'(x)dx$

And then,

$f(x) * g(x) = \int f'(x) * g(x)dx + \int f(x) * g'(x)dx$

Now, the substitution rule is used.

Usually, u and v are used as the functions we substitute f(x) and g(x):

If $u = f(x), than\ f'(x) = \frac{du}{dx}$

From here,

$du = f'(x)dx$

If $v = g(x), than\ g'(x) = \frac{dv}{dx}$

From here,

$dv = g'(x)dx$

$f(x) * g(x) = \int f'(x) * g(x)dx + \int f(x) * g'(x)dx$ becomes:

$uv = \int vdu + \int udv$

From here:

$\int \boldsymbol{udv} = \boldsymbol{uv} - \int \boldsymbol{vdu}$ <u>the integration by parts formula.</u>

EXAMPLE

If $f(x) = \frac{\ln(x)}{x^2}$ find $F(x)$

$u = \ln(x)\ so\ u' = \frac{du}{dx} = \frac{1}{x}$

$du = \frac{1}{x}dx$

$dv = \frac{1}{x^2}dx\ so\ v = \int \frac{1}{x^2}dx = \int x^{-2}dx = -x^{-1}$

By using the integration by parts formula, we have:

$\int udv = uv - \int vdu$

$\int \ln(x)\left(\frac{1}{x^2}dx\right) = \ln(x)\,(-x^{-1}) - \int(-x^{-1})\left(\frac{1}{x}dx\right) = -\frac{\ln(x)}{x} + \int(x^{-2}dx) = -\frac{\ln(x)}{x} - \frac{1}{x} = -\frac{1}{x}[\ln(x) + 1]$

Chapter 4. E. c. Methods of Integration – By parts

This mountain is the highest in Switzerland at a height of 4364 m. It is situated near the Swiss – Italian border. Its name is Monte ...

Determine which answer is correct. In the table at the bottom of the page, cross off all the letters for the correct answers The word that remains uncrossed is the answer.

1) If $f(x) = x^4 \ln(x)$ then $F(x) = \int f(x)dx = \frac{1}{5}x^5\left[\ln(x) - \frac{1}{5}\right] + C$

2) If $f(x) = x^3 cos(x)$ then $F(x) = \int f(x)dx = \sin(x)\,(x^3) + 3x^2\cos(x) + 6x sin(x) +$

$6\cos(x) + C$

3) If $f(x) = \frac{\ln(x)}{x^3}$ then $F(x) = \int f(x)dx = \frac{3}{x^2} + C$

4) If $f(x) = x^2 sin(x)$ then $F(x) = \int f(x)dx = 2x\sin(x) + 2\cos(x) + C$

5) If $f(x) = e^x cos(x)$ then $F(x) = \int f(x)dx = \frac{1}{2}e^x[\sin(x) + \cos(x)] + C$

6) If $f(x) = x^2 e^x$ then $F(x) = \int f(x)dx = \frac{1}{2}e^x 3x + C$

7) If $f(x) = \frac{\ln(x)}{x}$ then $F(x) = \int f(x)dx = \frac{1}{2}ln^2(x) + C$

8) If $f(x) = x\sqrt{1 - sin^2(x)}$ then $F(x) = \int f(x)dx = \cos(x) + C$

9) If $f(x) = xln(x)$ then $F(x) = \int f(x)dx = \frac{x^2}{2}\left[\ln(x) - \frac{1}{2}\right] + C$

10) If $f(x) = e^x \sin(2x)$ then $F(x) = \int f(x)dx = \frac{1}{5}sin(2x)e^x - \frac{2}{5}e^x cos(2x) + C$

1	2	3	4	5	6	7	8	9	10
C	A	R	O	L	S	U	A	I	M

4.F. Integration - Applications

a. Aria under a curve, volume of solids, average value of functions

Theory and Examples

Remember from 4D that the fundamental theorem of calculus says that:

$$\int_a^b f(x)dx = \int_0^b f(x)dx - \int_0^a f(x)dx = F(b) - F(a), where\ F'(x) = f(x)$$

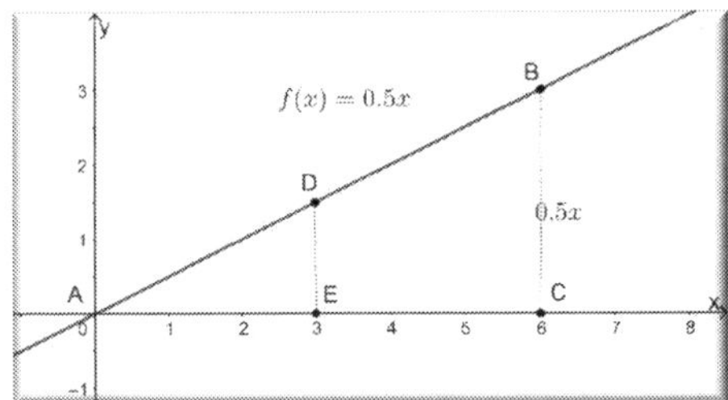

Before we explain the formula, we go back to the area problem in 4.B.

Area Problem extended

We want to calculate the area of EDBC. First, we calculate the area of the triangle ABC. The function that constitutes the hypotenuse is:

$f(x) = \frac{1}{2}x$

The base of the triangle AC is x.
The height CB is $f(x) = 0.5x$

$Area = \frac{1}{2}AC * CB = \left(\frac{1}{2}x\right)x = \frac{1}{4}x^2$

What is the derivative of $\frac{1}{4}x^2$?

$(\frac{1}{4}x^2)' = \frac{1}{4} * 2x = \frac{1}{2}x = f(x)$, so,

$\frac{1}{4}x^2 = F(x) = f'(x)$

Area of the surface ABC that is below y=f(x) in between zero and 6 is the definite integral

$\int_0^6 f(x)dx = F(x)$ between x=0 and x=6

$F(6) = \frac{1}{4}(6)^2 = \frac{36}{4} = 9$

$F(0) = 0$

Second, we calculate the area of the triangle ADE.
The function that constitutes the hypotenuse is:

$f(x) = \frac{1}{2}x$

Area of the surface ADE that is below y=f(x) in between zero and 3 is the definite integral

$\int_0^3 f(x)dx = F(x)$ between x=0 and x=3.

$F(3) = \frac{1}{4}(3)^2 = \frac{9}{4} = 2.25$

$Geometric\ Area_{EDBC} = \frac{(Big\ base+small\ base)*Height}{2} = \frac{(3+1.5)*3}{2} = 6.75$

What we notice is that if we subtract $F(6) - F(3) = 9 - 2.25 = 6.75$ equal with the area of EDBC

So, the area situated below the function $f(x) = \frac{1}{2}x$ between x=3 and x=6 is given by:

$\int_3^6 f(x)dx = \int_0^6 f(x)dx - \int_0^3 f(x)dx = F(6) - F(3) = 9 - 2.25 = 6.75, where\ F'(x) = f(x)$

EXAMPLE

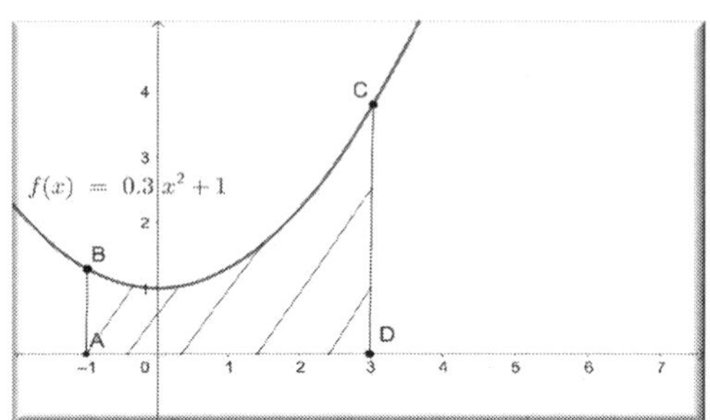

Find the area under the graph of the function $f(x) = 0.3x^2 + 1$ between x=-1 and x=3 using the Fundamental Theorem of Calculus.

$\int_{-1}^3 f(x)dx = F(3) - F(-1)$

$F(x) = \int f(x)dx = \frac{0.3}{3}x^3 + x + C$

So,

$F(3) = \frac{0.3}{3}(3)^3 + (3) + C = 5.7 + C$

$F(-1) = \frac{0.3}{3}(-1)^3 + (-1) + C = -1.1 + C$

So,

$\int_{-1}^3 f(x)dx = F(3) - F(-1) = 5.7 + C - (-1.1) - C = 6.8$

<u>Volume for solids problem</u>

Let us calculate the volume of a cone. The slant of the cone is the function y=f(x). This line will rotate around x axis. At any x value, y represents the radius of the circle at that x value.

Through the integral, we add all the surfaces with y=f(x) radius between the x initial and x final and obtain the volume of the cone.

EXAMPLE

Suppose we want to find the volume of a cone with the slant y=f(x)=x.

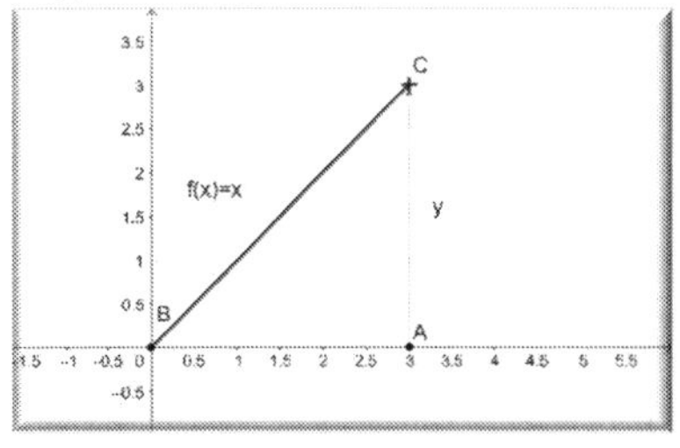

The slant rotates around x axis.

We consider y=f(x) as the radius of all the circles that constitute the cone.

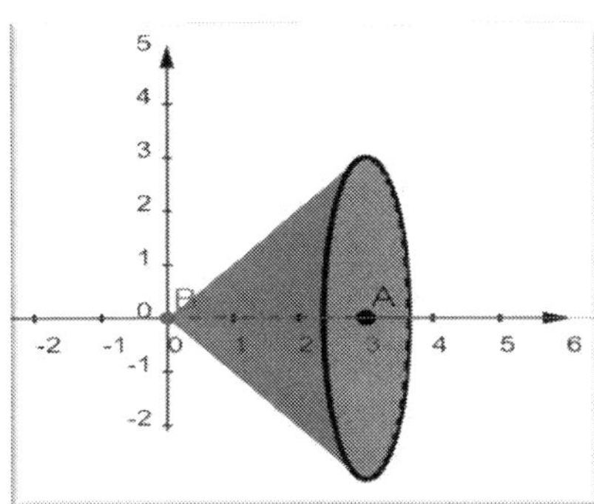

The area of the circle is given by the formula:

$A = \pi Radius^2 = \pi[f(x)]^2 = \pi x^2$

The sum of all the circles that is the volume of the cone is given by the definite integral:

$Volume = \int_0^3 \pi x^2 dx$

So,

$Volume = \int_0^3 \pi x^2 dx = \pi(3)^2 - \pi(0)^2 = 9\pi$

The average value of a function

The average value of a function on an interval [a,b] is calculated using the formula:

$Average\ value = \frac{1}{b-a}\int_a^b f(x)dx$

EXAMPLE

The average value of the function $f(x) = 2x^2 + \frac{1}{x}$ on the interval [2,6] is calculated:

$\int_2^6 f(x)dx = F(6) - F(2)$

First, we calculate the indefinite integral from f(x):

$F(x) = \int f(x)dx = \int(2x^2 + \frac{1}{x})dx = 2\int x^2 dx + \int \frac{1}{x} dx = 2\frac{1}{3}x^3 + \ln(x) + C$

Second, we calculate the value of F(x) for x=2 and x=6

$F(6) = 2\frac{1}{3}(6)^3 + \ln(6) = 144 + 1.79 = 145.79$

$F(2) = 2\frac{1}{3}(2)^3 + \ln(2) = 5.33 + 0.69 = 6.02$

We apply the Fundamental Theorem of Calculus.

$\int_2^6 f(x)dx = F(6) - F(2) = 145.79 - 6.02 = 139.77$

$Average\ value = \frac{1}{6-2}\int_2^6 f(x)dx = \frac{139.77}{4} = 34.94$

Chapter 4. F. a. Aria under a curve, volume of solids, average value of functions

On July 20th 1969, Neil Armstrong and Buzz Aldrin were the first humans that landed on the surface of the Moon. Armstrong noticed that the soil is "almost like a ...."

Determine which answer is correct. In the table at the bottom of the page, cross off all the letters for the correct answers. The word that remains uncrossed is the answer.

PAGE 1

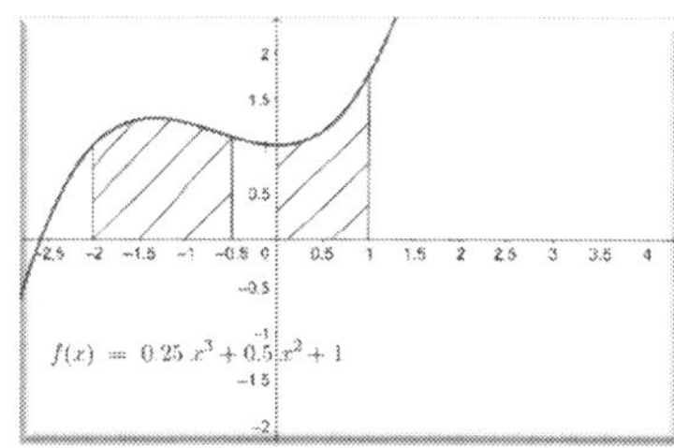

1) The area under $f(x) = \frac{1}{4}x^3 + \frac{1}{2}x^2 + 1$ between x=-2 and x=-0.5 is: 7

2) The area under $f(x) = \frac{1}{4}x^3 + \frac{1}{2}x^2 + 1$ between x=0 and x=1 is: 1.222

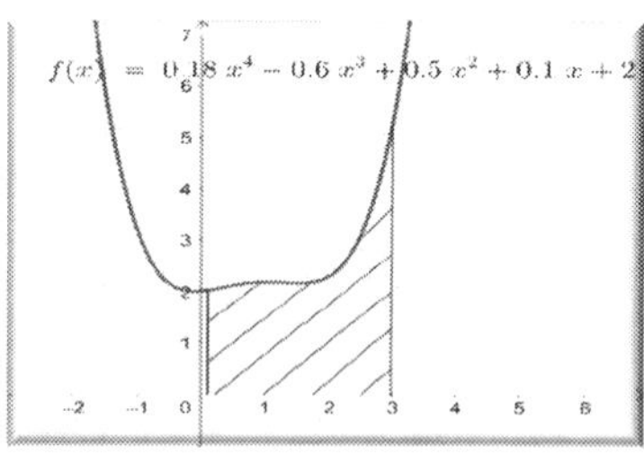

3) The area under $f(x) = 0.18x^4 - 0.6x^3 + 0.5x^2 + x + 2$ between x=0.1 and x=3 is:25

4) The volume of the object that is made by the function $f(x) = x^2 + 2$ that rotate around x axis between x=0 and x=2 is: V=10 π

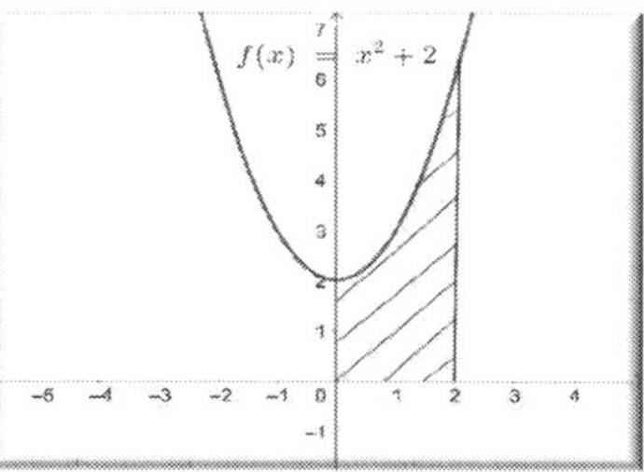

PAGE 2

5) The volume of the object that is made by the function $f(x) = -0.3x^2 + 4$ that rotate around x axis between x=0.5 and x=3 is: V= 22.87π

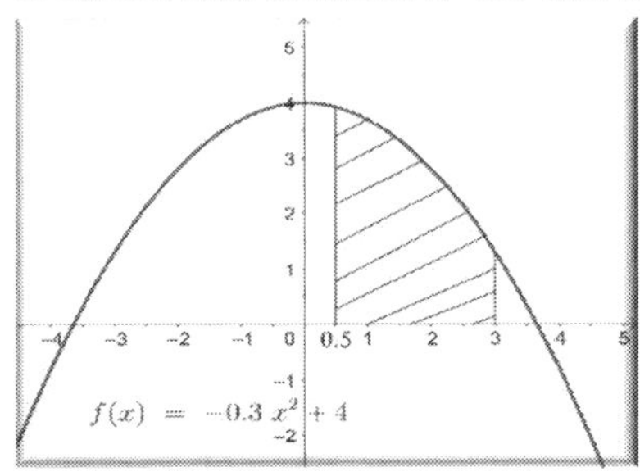

6) The volume of the sphere that is made by the function $4 = x^2 + y^2$ that rotate around x axis between x=-2 and x=2 is: V= 1.66π

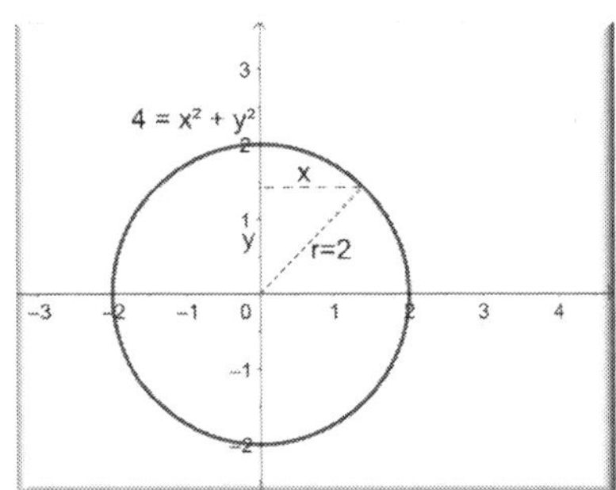

7) The volume of the object that is made by the function $f(x) = x^2$ that rotate around x axis between x=0 and x=3 is: 48.6π

8) The volume of the object that is made by the function $f(x) = x$ that rotate around x axis between x=1 and x=5 is: 2

9) The average value of the function $f(x) = 2x + \sin(x)$ on the interval [1,3] is: 4

10) The average value of the function $f(x) = x^2 + \frac{1}{x}$ on the interval [1,5] is: 10

1	2	3	4	5	6	7	8	9	10
P	A	O	W	I	D	L	E	U	R

4.F. Integration

b. Differential equations, Initial value problems, Slope fields

Theory and Examples

A differential equation is an equation where there are functions and their derivatives.

EXAMPLE

If y=f(x)=3x+1 solve the differential equation.

$y' + 2y - 3 = 0$

First, we calculate the derivative of y:

$(3x + 1)' = 3$

Second, we substitute y=3x+1 and the derivative of y in the equation.

$3 + 2(3x + 1) + 1 = 0$

Third, we solve the equation for x.

$3 + 6x + 2 + 1 = 0$

$6x + 6 = 0\ so, x = -1$

Initial value problems are differential equations that have given an initial value. In physics models many times the model starts at an initial value.

EXAMPLE

If the differential equation is $y' = \frac{dy}{dx} = x^2 + 3x$, the initial condition is y(1)=2, find y:

$dy = (x^2 + 3x)dx$

We integrate both sides;

$\int dy = \int (x^2 + 3x)dx = \int x^2 dx + \int 3x dx = \frac{1}{3}x^3 + \frac{3}{2}x^2 + C$

$y = \frac{1}{3}x^3 + \frac{3}{2}x^2 + C$

We plug y=2 and x=1 in the formula to find C:

$2 = \frac{1}{3}(1)^3 + \frac{3}{2}(1)^2 + C$

$2 - \frac{1}{3} - \frac{3}{2} = C$

$\frac{12}{6} - \frac{2}{6} - \frac{9}{6} = \frac{1}{6} = 0.167 = C$

We write the final form of the function y with the constant C=0.167.

$y = \frac{1}{3}x^3 + \frac{3}{2}x^2 + 0.167$

Slope fields are all the antiderivative functions F(x) that we obtain for any constant C.

EXAMPLE

If one of the functions whose derivative is $f(x) = e^x$ is:

$F(x) = \int e^x dx = e^x + C$ So, one of the functions is $F(x) = e^x + 3$

Chapter 4. F. b. Differential equations, Initial value problems, Slope fields

Only the members of the Roman senate were allowed to wear a tunic called toga with a broad -purple stripe. This purple stripe was called *latus...*

Determine which answer is correct. In the table at the bottom of the page, cross off all the letters of the correct answers. The word that remains uncrossed is the answer.

The trajectory of the Space Shuttle in the first minutes is represented by:

$h(t) = 2008 - 0.047t^3 + 18.3t^2 - 345t$

1) The velocity of the Space Shuttle at 20 seconds is: 330.6 m/s

2) The acceleration of the Space Shuttle at 20 seconds is: $3\ m/s^2$

A formula one speed car has the acceleration formula: a(t)=t+4 m/s^2

3) The velocity of the car after 10 seconds is: $90\ m/s$

4) The distance traveled by the car after 10 seconds is: $66.6\ m$

5) If the differential equation is $y' = \frac{dy}{dx} = 15 + x$, the initial condition is y(0)=3, then $y = 15x + \frac{1}{2}x^2 + 3$

6) If the differential equation is $y' = \frac{dy}{dx} = 2x^2 - 3x$, the initial condition is y(1)=2, then $y = \frac{2}{3}x^3 - \frac{3}{2}x^2 + 1.83$

7) If the differential equation is $y' = \frac{dy}{dx} = e^x + 4x$, the initial condition is y(0)=5, then $y = e^x + 2x^2 + 40$

8) One of the functions whose derivative is $f(x) = 4x$ is $F(x) = 2x^2 + 37$

9) One of the functions whose derivative is $f(x) = 3x^{-1}$ is $F(x) = 3x + 3$

10) One of the functions whose derivative is $f(x) = e^x$ is $F(x) = \frac{5e^{2x}}{2} + 3$

1	2	3	4	5	6	7	8	9	10
E	C	I	L	T	A	V	O	U	S

QUICK ANSWERS

Chapter 4

4.A GAUDI

4.B GROUSE

4.C FROG

4.D DNOCES (SECOND)

4.E.a OSTIA

4.E.b WRIGHT

4.E.c ROSA

4.F.a POWDER

4.F.b CLAVUS

FULL SOLUTIONS

CHAPTER 4

Chapter 4. A. Definition of an integral and notation

1. Correct

An integral is a weighted sum of the values of the function times the infinitesimal widths dx.

2. Incorrect.

The number of infinitesimal widths has to be as high as it is possible not necessarily 100.

3. Correct

The notation of definite integral in the interval [a,b] is: $\int_a^b f(x)dx$.

4. Incorrect

The sign $\int$ represents the sum of products between $f(x)$ and dx.

5. Incorrect

The notation of indefinite integral or antiderivative is: $\int f(x)dx$.

6. Incorrect

The function $f(x)$ under the integral sign it is called integrand.

7. Correct

The first documented technique that tried to calculate the integral was used by the ancient Greek astronomer Euxodus around 370 BC.

8. Incorrect

The Monte Carlo method is used in statistics and probabilities to generate random numbers that that follow certain criteria and calculating the average of these numbers.

Chapter 4. B. Definite and indefinite integrals

1. Correct

If $f(x) = 3x^2 - e^x$, then $\int f(x)dx = x^3 - e^x + C$

$\int f(x)dx = \int(3x^2 - e^x)dx = \int(3x^2)dx - \int e^x dx = \frac{3x^{2+1}}{3} - e^x = x^3 - e^x + C$

2. Incorrect

If $f(x) = 2x^4 + \ln(x)$, then $\int f(x)dx = \frac{2}{5}x^5 + xln(x) - x + C$

$\int f(x)dx = \int[2x^4 + \ln(x)]dx = \int 2x^4\, dx + \int[\ln(x)]dx = \frac{2}{5}x^5 + xln(x) - x + C$

3. Incorrect

If $f(x) = \ln(x) - 2^x$, then $\int f(x)dx = xln(x) - x - \frac{2^x}{\ln(2)} + C$

$\int f(x)dx = \int[\ln(x)]dx - \int 2^x dx = xln(x) - x - \frac{2^x}{\ln(2)} + C$

4. Correct

If $f(x) = 1 + \ln(x)$, then $\int f(x)dx = xln(x) + C$

$\int f(x)dx = \int dx + \int[\ln(x)\, dx = x + xln(x) - x = xln(x) + C$

5. Incorrect

If $f(x) = 2 - cos\,(x)$, then $\int f(x)dx = 2x - \sin(x) + C$

$\int f(x)dx = \int 2dx - \int cos(x)dx = 2x - \sin(x) + C$

6. Correct

If $f(x) = x^2 - 2x$, then $\int_0^x f(x)dx = \frac{1}{3}x^3 - x^2$

$\int_0^x f(x)dx = \int_0^x x^2dx - \int_0^x 2xdx = \frac{1}{3}x^3 - x^2$

7. Incorrect

If $f(x) = 2x^2 + 3x - 4$, then $\int_0^x f(x)dx = -1\frac{5}{6}$

$\int_0^1 f(x) = \int_0^1(2x^2 + 3x - 4)dx = \int_0^1 2x^2dx + \int_0^1 3xdx - \int_0^1 4dx = \frac{2}{3}(1)^3 + \frac{3}{2}(1)^2 - -4(1) = \frac{2}{3} + \frac{3}{2} - 4 = \frac{4}{6} + \frac{9}{6} - \frac{24}{6} = -\frac{11}{6} = -1\frac{5}{6}$

8. Incorrect

If $f(x) = x^2 - 4\cos(x)$, then $\int_0^6 f(x)dx = 71.58$

$\int_0^6 f(x)dx = \int_0^6[x^2 - 4\cos(x)]dx = \int_0^6 x^2dx - 4\int_0^6 \cos(x)\, dx = \frac{1}{3}(6)^3 - 4\sin(6) = 72 - 4(0.104) = 71.58$

9. Correct

If $f(x) = \cos(x) + 3x$, then $\int_0^4 f(x)dx = 24.069$

$\int_0^4 f(x)dx = \int_0^4 [\cos(x) + 3x]dx = \int_0^4 \cos(x)dx + 3\int_0^4 xdx = \sin(4) + \frac{3}{2}(4)^2 = 0.069 + 24 = 24.069$

10. Incorrect

If $f(x) = 2x + \sqrt{x}$, then $\int_0^3 f(x)dx = 9(1 + \frac{\sqrt{3}}{2})$

$\sqrt{x} = x^{\frac{1}{2}}\ so, \int x^{\frac{1}{2}}dx = \frac{3x^{\frac{3}{2}}}{2}$

$\int_0^3 f(x)dx = \int_0^3 (2x + \sqrt{x})dx = \int_0^3 2xdx + \int_0^3 \sqrt{x}dx = (3)^2 + \frac{3(3)^{\frac{3}{2}}}{2} = 9 + \frac{9\sqrt{3}}{2} = 9(1 + \frac{\sqrt{3}}{2})$

Chapter 4. C. Approximations-Riemann sum, rectangle method, trapezoidal method

1. Correct

If $f(x) = 2x^2 - 3x$, using Riemann left sum, the integral $\int_{-2}^{3} f(x)dx = 20$

The interval $\Delta x = 1$

$\int_{-2}^{3} f(x)dx \cong \sum_{-2}^{3} f(x)\Delta x = f(-2)\Delta x + f(-1)\Delta x + f(0)\Delta x + f(1)\Delta x + f(2)\Delta x$

$f(-2) = 2(-2)^2 - 3(-2) = 8 + 6 = 14$

$f(-1) = 2(-1)^2 - 3(-1) = 2 + 3 = 5$

$f(0) = 2(0)^2 - 3(0) = 0$

$f(1) = 2(1)^2 - 3(1) = 2 - 3 = -1$

$f(2) = 2(2)^2 - 3(2) = 8 - 6 = 2$

$\sum_{-2}^{3} f(x)\Delta x = f(-2)\Delta x + f(-1)\Delta x + f(0)\Delta x + f(1)\Delta x + f(2)\Delta x = 14 + 5 - 1 + 2 = 20$

$\int_{-2}^{3} f(x)dx \cong 20$

2. Incorrect

If $f(x) = 4x^3 + 3x^2 - 2x + 1$, using Riemann left sum, the integral $\int_{-3}^{4} f(x)dx \cong 91$

The interval $\Delta x = 1$

$\int_{-3}^{4} f(x)dx \cong \sum_{-3}^{4} f(x)\Delta x = f(-3)\Delta x + f(-2)\Delta x + f(-1)\Delta x + f(0)\Delta x + f(1)\Delta x + f(2)\Delta x + f(3)\Delta x$

$f(-3) = 4(-3)^3 + 3(-3)^2 - 2(-3) + 1 = 4(-27) + 27 + 6 + 1 = -81 + 7 = -74$

$f(-2) = 4(-2)^3 + 3(-2)^2 - 2(-2) + 1 = -32 + 12 + 4 + 1 = -15$

$f(-1) = 4(-1)^3 + 3(-1)^2 - 2(-1) + 1 = -4 + 3 + 2 + 1 = 2$

$f(0) = 4(0)^3 + 3(0)^2 - 2(0) + 1 = 1$

$f(1) = 4(1)^3 + 3(1)^2 - 2(1) + 1 = 4 + 3 - 2 + 1 = 6$

$f(2) = 4(2)^3 + 3(2)^2 - 2(2) + 1 = 32 + 12 - 4 + 1 = 41$

$f(3) = 4(3)^3 + 3(3)^2 - 2(3) + 1 = 4(27) + 27 - 6 + 1 = 130$

$\int_{-3}^{4} f(x)dx \cong \sum_{-3}^{4} f(x)\Delta x = f(-3)\Delta x + f(-2)\Delta x + f(-2)\Delta x + f(0)\Delta x + f(1)\Delta x + f(2)\Delta x + f(3)\Delta x =$
$-74 - 15 + 2 + 1 + 6 + 41 + 130 = 91$

3. Correct

If $f(x) = 3x^3 - 2x^2 + x - 1$, using Rectangle Method, the integral $\int_{-2}^{3} f(x)dx \cong 21.875$

The interval $\Delta x = 1$

The mid points are: x=-1.5; x=-0.5; x=0.5; x=1.5; x=2.5

$\int_{-2}^{3} f(x)dx \cong \sum_{-2}^{3} f(x)\Delta x = f(-1.5)\Delta x + f(-0.5)\Delta x + f(0.5)\Delta x + f(1.5)\Delta x + f(2.5)\Delta x$

$f(-1.5) = 3(-1.5)^3 - 2(-1.5)^2 + (-1.5) - 1 = 3(-3.375) - 2(2.25) - 1.5 - 1 = -17.125$

$f(-0.5) = 3(-0.5)^3 - 2(-0.5)^2 + (-0.5) - 1 = 3(-0.125) - 2(0.25) - 0.5 - 1 = -2.375$

$f(0.5) = 3(0.5)^3 - 2(0.5)^2 + (0.5) - 1 = 3(0.125) - 2(0.25) + 0.5 - 1 = -0.625$

$f(1.5) = 3(1.5)^3 - 2(1.5)^2 + (1.5) - 1 = 3(3.375) - 2(2.25) + 1.5 - 1 = 6.125$

$f(2.5) = 3(2.5)^3 - 2(2.5)^2 + (2.5) - 1 = 3(15.625) - 2(6.25) + 2.5 - 1 = 35.875$

$\int_{-2}^{3} f(x)dx = \sum_{-2}^{3} f(x)\Delta x = f(-1.5)\Delta x + f(-0.5)\Delta x + f(0.5)\Delta x + f(1.5)\Delta x + f(2.5)\Delta x = -17.125 -$
$2.375 - 0.625 + 6.125 + 35.875 = 21.875$

4. Incorrect

If $f(x) = x^2 + 2x + \sqrt[3]{x}$, using Rectangle Method, the integral $\int_{-3}^{3} f(x)dx \cong 17.5$

The interval $\Delta x = 1$

The mid points are: x=-2.5; x=-1.5; x=-0.5; x=0.5; x=1.5; x=2.5

$\int_{-3}^{3} f(x)dx \cong \sum_{-3}^{3} f(x)\Delta x = f(-2.5)\Delta x + f(-1.5)\Delta x + f(-0.5)\Delta x + f(0.5)\Delta x + f(1.5)\Delta x + f(2.5)\Delta x$

$f(-2.5) = (-2.5)^2 + 2(-2.5) + \sqrt[3]{-2.5} = 6.25 - 5 - 1.35 = -0.1$

$f(-1.5) = (-1.5)^2 + 2(-1.5) + \sqrt[3]{-1.5} = 2.25 - 3 - 1.14 = -1.89$

$f(-0.5) = (-0.5)^2 + 2(-0.5) + \sqrt[3]{-0.5} = 0.25 - 1 - 0.79 = -1.54$

$f(0.5) = (0.5)^2 + 2(0.5) + \sqrt[3]{0.5} = 0.25 + 1 + 0.79 = 2.04$

$f(1.5) = (1.5)^2 + 2(1.5) + \sqrt[3]{1.5} = 2.25 + 3 + 1.14 = 6.39$

$f(2.5) = (2.5)^2 + 2(2.5) + \sqrt[3]{2.5} = 6.25 + 5 + 1.35 = 12.6$

$\int_{-3}^{3} f(x)dx \cong \sum_{-3}^{3} f(x)\Delta x = f(-2.5)\Delta x + f(-1.5)\Delta x + f(-0.5)\Delta x + f(0.5)\Delta x + f(1.5)\Delta x + f(2.5)\Delta x =$
$-0.1 - 1.89 - 1.54 + 2.04 + 6.39 + 12.6 = 17.5$

5. Correct

If $f(x) = 4x^3 + \sin(x)$, using Riemann left sum, the integral $\int_{-3}^{3} f(x)dx \cong -108.05$

The interval $\Delta x = 1$

$\int_{-3}^{3} f(x)dx \cong \sum_{-3}^{3} f(x)\Delta x = f(-3)\Delta x + f(-2)\Delta x + f(-1)\Delta x + f(0)\Delta x + f(1)\Delta x + f(2)\Delta x$

$f(-3) = 4(-3)^3 + \sin(-3) = 4(-27) - 0.05 = -108 - 0.05 = -108.05$

$f(-2) = 4(-2)^3 + \sin(-2) = -32 - 0.03 = -32.03$

$f(-1) = 4(-1)^3 + \sin(-1) = -4 - 0.01 = -4.01$

$f(0) = 0$

$f(1) = 4(1)^3 + \sin(1) = 4 + 0.01 = 4.01$

$f(2) = 4(2)^3 + \sin(2) = 32 + 0.05 = 32.05$

$\int_{-3}^{3} f(x)dx \cong \sum_{-3}^{3} f(x)\Delta x = f(-3)\Delta x + f(-2)\Delta x + f(-1)\Delta x + f(0)\Delta x + f(1)\Delta x + f(2)\Delta x = -108.050$

6. Incorrect

If $f(x) = 3x^3 - 2x^2 + 1$, using left side Riemann sum, the integral $\int_{-2}^{2} f(x)dx \cong -32$

The interval $\Delta x = 1$

The points are: x=-2; x=-1; x=0; x=1;

$\int_{-2}^{2} f(x)dx \cong \sum_{-2}^{2} f(x)\Delta x = f(-2)\Delta x + f(-1)\Delta x + f(0)\Delta x + f(1)\Delta x$

$f(-2) = 3(-2)^3 - 2(-2)^2 + 1 = -24 - 8 + 1 = -31$

$f(-1) = 3(-1)^3 - 2(-1)^2 + 1 = -3 - 2 + 1 = -4$

$f(0) = 3(0)^3 - 2(0)^2 + 1 = 1$

$f(1) = 3(1)^3 - 2(1)^2 + 1 = 3 - 2 + 1 = 2$

$\int_{-2}^{2} f(x)dx \cong \sum_{-2}^{2} f(x)\Delta x = f(-2)\Delta x + f(-1)\Delta x + f(0)\Delta x + f(1)\Delta x = -31 - 4 + 1 + 2 = -32$

7. Correct

If $f(x) = 3x^3 - 2x^2 + 1$, using left side Riemann sum, the integral $\int_{-2}^{2} f(x)dx \cong -19$

The interval $\Delta x = 0.5$

The points are: x=-2; x=-1.5; x=-1; x=-0.5; x=0; x=0.5; x=1; x=1.5

$\int_{-2}^{2} f(x)dx \cong \sum_{-2}^{2} f(x)\Delta x = f(-2)\Delta x + f(-1.5)\Delta x + f(-1)\Delta x + f(-0.5)\Delta x + f(0)\Delta x + f(0.5)\Delta x +$
$f(1)\Delta x + f(1.5)\Delta x$

$f(-2) = 3(-2)^3 - 2(-2)^2 + 1 = -24 - 4 + 1 = -31$

$f(-1.5) = 3(-1.5)^3 - 2(-1.5)^2 + 1 = 3(-3.375) - 4.5 + 1 = -10.125 - 4.5 + 1 = -13.625$

$f(-1) = 3(-1)^3 - 2(-1)^2 + 1 = -3 - 2 + 1 = -4$

$f(-0.5) = 3(-0.5)^3 - 2(-0.5)^2 + 1 = 3(-0.125) - 0.5 + 1 = 0.125$

$f(0) = 3(0)^3 - 2(0)^2 + 1 = 1$

$f(0.5) = 3(0.5)^3 - 2(0.5)^2 + 1 = 3(0.125) - 0.5 + 1 = 0.875$

$f(1) = 3(1)^3 - 2(1)^2 + 1 = 3 - 2 + 1 = 2$

$f(1.5) = 3(1.5)^3 - 2(1.5)^2 + 1 = 3(3.375) - 4.5 + 1 = 10.125 - 4.5 + 1 = 6.625$

$\int_{-2}^{2} f(x)dx \cong \sum_{-2}^{2} f(x)\Delta x = f(-2)(0.5) + f(-1.5)(0.5) + f(-1)(0.5) + f(-0.5)(0.5) + f(0)(0.5) + f(0.5)(0.5) + f(1)(0.5) + f(1.5)(0.5) = (-31)(0.5) + (-13.625)(0.5) + (-4)(0.5) + (0.125)(0.5) + 1 * (0.5) + (0.875)(0.5) + 2 * (0.5) + (6.625)(0.5) = -15.5 - 6.812 - 2 - 0.0625 + 0.5 + 0.437 + 1 + 3.312 = -19$

8. Incorrect

If $f(x) = 3x^3 - 2x^2 + 1$, using trapezoidal method, the integral $\int_{-2}^{2} f(x)dx \cong -6$

The interval $\Delta x = 1$

$\int_{-2}^{2} f(x)dx \cong \sum_{i=-2}^{i=2} \frac{f(x_i)+f(x_{i+1})}{2} \Delta x_i = \frac{f(-2)+f(-1)}{2} * 1 + \frac{f(-1)+f(0)}{2} * 1 + \frac{f(0)+f(1)}{2} * 1 + \frac{f(1)+f(2)}{2} * 1$

$f(-2) = 3(-2)^3 - 2(-2)^2 + 1 = -24 - 8 + 1 = -31$

$f(-1) = 3(-1)^3 - 2(-1)^2 + 1 = -3 - 2 + 1 = -4$

$f(0) = 3(0)^3 - 2(0)^2 + 1 = 1$

$f(1) = 3(1)^3 - 2(1)^2 + 1 = 3 - 2 + 1 = 2$

$f(2) = 3(2)^3 - 2(2)^2 + 1 = 24 - 8 + 1 = 17$

$\int_{-2}^{2} f(x)dx \cong \sum_{i=-2}^{i=2} \frac{f(x_i)+f(x_{i+1})}{2} \Delta x_i = \frac{-31-4}{2} * 1 + \frac{-4+1}{2} * 1 + \frac{1+2}{2} * 1 + \frac{2+17}{2} * 1 = -17.5 - 1.5 + 1.5 + 9.5 = -8$

9. Correct

If $f(x) = 3x^3 - 2x^2 + 1$, using trapezoidal method, the integral $\int_{-2}^{2} f(x)dx \cong -7$

The interval $\Delta x = 0.5$

$\int_{-2}^{2} f(x)dx \cong \sum_{i=-2}^{i=2} \frac{f(x_i)+f(x_{i+1})}{2} \Delta x_i = \frac{f(-2)+f(-1.5)}{2} * 0.5 + \frac{f(-1.5)+f(-1)}{2} * 0.5 + \frac{f(-1)+f(-0.5)}{2} * 0.5 + \frac{f(-0.5)+f(0)}{2} * 0.5 + \frac{f(0)+f(0.5)}{2} * 0.5 + \frac{f(0.5)+f(1)}{2} * 0.5 + \frac{f(1)+f(1.5)}{2} * 0.5 + \frac{f(1.5)+f(2)}{2} * 0.5$

$f(-2) = 3(-2)^3 - 2(-2)^2 + 1 = -24 - 8 + 1 = -31$

$f(-1.5) = 3(-1.5)^3 - 2(-1.5)^2 + 1 = 3(-3.375) - 2(2.25) + 1 = -10.125 - 4.5 + 1 = -13.625$

$f(-1) = 3(-1)^3 - 2(-1)^2 + 1 = -3 - 2 + 1 = -4$

$f(-0.5) = 3(-0.5)^3 - 2(-0.5)^2 + 1 = 3(-0.125) - 2(0.25) + 1 = -0.375 - 0.5 + 1 = 0.125$

$f(0) = 3(0)^3 - 2(0)^2 + 1 = 1$

$f(0.5) = 3(0.5)^3 - 2(0.5)^2 + 1 = 3(0.125) - 2(0.25) + 1 = 0.375 - 0.5 + 1 = 0.875$

$f(1) = 3(1)^3 - 2(1)^2 + 1 = 3 - 2 + 1 = 2$

$f(1.5) = 3(1.5)^3 - 2(1.5)^2 + 1 = 3(3.375) - 2(2.25) + 1 = 10.125 - 4.5 + 1 = 6.625$

$f(2) = 3(2)^3 - 2(2)^2 + 1 = 24 - 8 + 1 = 17$

$\int_{-2}^{2} f(x)dx \cong \sum_{i=-2}^{i=2} \frac{f(x_i)+f(x_{i+1})}{2} \Delta x_i = \frac{f(-2)+f(-1.5)}{2} * 0.5 + \frac{f(-1.5)+f(-1)}{2} * 0.5 + \frac{f(-1)+f(-0.5)}{2} * 0.5 + \frac{f(-0.5)+f(0)}{2} * 0.5 + \frac{f(0)+f(0.5)}{2} * 0.5 + \frac{f(0.5)+f(1)}{2} * 0.5 + \frac{f(1)+f(1.5)}{2} * 0.5 + \frac{f(1.5)+f(2)}{2} * 0.5 = \frac{-31-13.625}{2} * 0.5 +$

$\frac{-13.625-4}{2} * 0.5 + \frac{-4-0.125}{2} * 0.5 + \frac{0.125+1}{2} * 0.5 + \frac{1+0.875}{2} * 0.5 + \frac{0.875+2}{2} * 0.5 + \frac{2+6.625}{2} * 0.5 + \frac{6.625+17}{2} * 0.5 =$
$-11.16 - 4.41 - 0.97 + 0.28 + 0.468 + 0.72 + 2.156 + 5.91 = -7$

10. Correct

If $f(x) = x^2 - x - 6$, using trapezoidal method, the integral $\int_1^5 f(x)dx \cong 5.50$

The interval $\Delta x = 0.5$

$\int_1^5 f(x)dx = \sum_{i=1}^{i=4} \frac{f(x_i)+f(x_{i+1})}{2} \Delta x_i = \frac{f(1)+f(1.5)}{2} * 0.5 + \frac{f(1.5)+f(2)}{2} * 0.5 + \frac{f(2)+f(2.5)}{2} * 0.5 + \frac{f(2.5)+f(3)}{2} * 0.5 +$
$\frac{f(3)+f(3.5)}{2} * 0.5 + \frac{f(3.5)+f(4)}{2} * 0.5 + \frac{f(4)+f(4.5)}{2} * 0.5 + \frac{f(4.5)+f(5)}{2} * 0.5$

$f(1) = (1)^2 - 1 - 6 = -6$

$f(1.5) = (1.5)^2 - 1.5 - 6 = 2.25 - 1.5 - 6 = -5.25$

$f(2) = (2)^2 - 2 - 6 = 4 - 2 - 6 = -4$

$f(2.5) = (2.5)^2 - 2.5 - 6 = 6.25 - 2.5 - 6 = -2.25$

$f(3) = (3)^2 - 3 - 6 = 9 - 3 - 6 = 0$

$f(3.5) = (3.5)^2 - 3.5 - 6 = 12.25 - 3.5 - 6 = 2.75$

$f(4) = (4)^2 - 4 - 6 = 16 - 4 - 6 = 6$

$f(4.5) = (4.5)^2 - 4.5 - 6 = 20.25 - 4.5 - 6 = 9.75$

$f(5) = (5)^2 - 5 - 6 = 25 - 5 - 6 = 14$

$\int_1^5 f(x)dx = \sum_{i=1}^{i=4} \frac{f(x_i)+f(x_{i+1})}{2} \Delta x_i = \frac{f(1)+f(1.5)}{2} * 0.5 + \frac{f(1.5)+f(2)}{2} * 0.5 + \frac{f(2)+f(2.5)}{2} * 0.5 + \frac{f(2.5)+f(3)}{2} * 0.5 +$
$\frac{f(3)+f(3.5)}{2} * 0.5 + \frac{f(3.5)+f(4)}{2} * 0.5 + \frac{f(4)+f(4.5)}{2} * 0.5 + \frac{f(4.5)+f(5)}{2} * 0.5 = \frac{-6-5.25}{2} * 0.5 + \frac{-5.25-4}{2} * 0.5 +$
$\frac{-4-2.25}{2} * 0.5 + \frac{-2.25+0}{2} * 0.5 + \frac{0+2.75}{2} * 0.5 + \frac{2.75+6}{2} * 0.5 + \frac{6+9.75}{2} * 0.5 + \frac{9.75+14}{2} * 0.5 = -2.81 - 2.31 -$
$1.56 - 0.56 + 0.69 + 2.19 + 3.94 + 5.94 = 5.50$

Chapter 4. D. Fundamental Theorem of Calculus

1. Correct

If $f(x) = 3x^2 + 3x - 2$ then, the $\int_{-4}^5 f(x)dx = 184.5$

First, the indefinite integral of $f(x)$ is:

$\int f(x)dx = \int(3x^2 + 3x - 2)dx = \int 3x^2dx + \int 3xdx - \int 2\,dx = x^3 + \frac{3}{2}x^2 - 2x + C = F(x)$

Applying the Fundamental Theorem of Calculus, we have:

$\int_{-4}^5 f(x)dx = F(5) - F(-4)$

$F(5) = (5)^3 + \frac{3}{2}(5)^2 - 2(5) = 125 + 1.5 * 25 - 10 = 152.5 + C$

$F(-4) = (-4)^3 + \frac{3}{2}(-4)^2 - 2(-4) = -64 + 24 + 8 = -32 + C$

$\int_{-4}^{5} f(x)dx = F(5) - F(-4) = 152.5 - (-32) = 184.5$

2. Incorrect

If $f(x) = x^2 + 3\sqrt{x}$ then, the $\int_{1}^{5} f(x)dx = 102.41$

$\int \sqrt{x}dx = \int x^{\frac{1}{2}}dx = \frac{x^{\frac{1}{2}+1}}{\frac{1}{2}+1} + C = \frac{x^{\frac{3}{2}}}{\frac{3}{2}} + C = \frac{2}{3}\sqrt{x^3} + C = \frac{2x\sqrt{x}}{3} + C$

First, the indefinite integral of $f(x)$ is:

$\int f(x)dx = \int(x^2 + 3\sqrt{x})dx = \int x^2dx + \int 3\sqrt{x}dx = \frac{1}{3}x^3 + 3\frac{2x\sqrt{x}}{3} + C = \frac{1}{3}x^3 + 6\,x\sqrt{x} + C = F(x)$

Applying the Fundamental Theorem of Calculus, we have:

$\int_{1}^{5} f(x)dx = F(5) - F(1)$

$F(5) = \frac{1}{3}x^3 + 6\,x\sqrt{x} = \frac{1}{3}(5)^3 + 6 * 5\sqrt{5} = 41.66 + 30 * 2.23 = 108.74$

$F(1) = \frac{1}{3}x^3 + 6x\sqrt{x} = \frac{1}{3}(1)^3 + 6\,x\sqrt{x} = 0.33 + 6 = 6.83$

$\int_{1}^{5} f(x)dx = F(5) - F(1) = 108.74 - 6.83 = 102.41$

3. Incorrect

If $f(x) = 4x - 3$ then, the $\int_{-4}^{5} f(x)dx = -9$

First, the indefinite integral of $f(x)$ is:

$\int f(x)dx = \int(4x - 3)dx = \int 4xdx - \int 3dx = \frac{4x^2}{2} - 3x + C = 2\,x^2 - 3x + C = F(x)$

Applying the Fundamental Theorem of Calculus, we have:

$\int_{-4}^{5} f(x)dx = F(5) - F(-4)$

$F(5) = 2(5)^2 - 3(5) = 50 - 15 = 35$

$F(-4) = 2(-4)^2 - 3(-4) = 32 + 12 = 44$

$\int_{-4}^{5} f(x)dx = F(5) - F(-4) = 35 - 44 = -9$

4. Correct

If $f(x) = 5x + 6\sqrt{x}$ then, the $\int_{2}^{7} f(x)dx = 175.3$

$\int \sqrt{x}dx = \int x^{\frac{1}{2}}dx = \frac{x^{\frac{1}{2}+1}}{\frac{1}{2}+1} + C = \frac{x^{\frac{3}{2}}}{\frac{3}{2}} + C = \frac{2}{3}\sqrt{x^3} + C = \frac{2x\sqrt{x}}{3} + C$

First, the indefinite integral of $f(x)$ is:

$\int f(x)dx = \int(5x + 6\sqrt{x})dx = \int 5xdx + \int 6\sqrt{x}dx = \frac{5x^2}{2} + 6\frac{2x\sqrt{x}}{3} + C = \frac{5x^2}{2} + 4x\sqrt{x} + C = F(x)$

Applying the Fundamental Theorem of Calculus, we have:

$\int_{2}^{7} f(x)dx = F(7) - F(2)$

$F(7) = \frac{5(7)^2}{2} + 4(7)\sqrt{7} = 122.5 + 28\sqrt{7} = 122.5 + 74.08 = 196.58$

$F(2) = \frac{5(2)^2}{2} + 4(2)\sqrt{2} = 10 + 8(1.41) = 21.28$

$\int_2^7 f(x)dx = F(7) - F(2) = 196.58 - 21.28 = 175.3$

5. Incorrect

If $f(x) = \frac{7}{x} + x - 3$ then, the $\int_1^5 f(x)dx = 11.26$

First, the indefinite integral of $f(x)$ is:

$\int f(x)dx = \int \left(\frac{7}{x} + x - 3\right) dx = \int \frac{7}{x}dx + \int xdx - \int 3dx = 7lnx + \frac{x^2}{2} - 3x + C = F(x)$

Applying the Fundamental Theorem of Calculus, we have:

$\int_1^5 f(x)dx = F(5) - F(1)$

$F(5) = 7ln(5) + \frac{(5)^2}{2} - 3(5) = 7(1.6) + 12.5 - 15 = 11.2 + 12.5 - 15 = 8.76$

$F(1) = 7ln(1) + \frac{(1)^2}{2} - 3(1) = 0 + 0.5 - 3 = -2.5$

$\int_1^5 f(x)dx = F(5) - F(1) = 8.76 - (-2.5) = 11.26$

6. Incorrect

If $f(x) = \frac{4+x^2}{x} - 3x + 7$ then, the $\int_2^8 f(x)dx = -12.49$

$f(x) = \frac{4+x^2}{x} - 3x + 7 = \frac{4}{x} + x - 3x + 7 = \frac{4}{x} - 2x + 7$

First, the indefinite integral of $f(x)$ is:

$\int f(x)dx = \int \left(\frac{4}{x} - 2x + 7\right) dx = \int \frac{4}{x}dx - \int 2xdx + \int 7dx = 4lnx - x^2 + 7x + C = F(x)$

Applying the Fundamental Theorem of Calculus, we have:

$\int_2^8 f(x)dx = F(8) - F(2)$

$F(8) = 4\ln(8) - (8)^2 + 7(8) = 4(2.07) - 64 + 56 = 0.31$

$F(2) = 4\ln(2) - (2)^2 + 7(2) = 4(0.69) - 4 + 14 = 12.77$

$\int_2^8 f(x)dx = F(8) - F(2) = 0.28 - 12.77 = -12.49$

7. Correct

If $f(x) = 4x + e^x$ then, the $\int_{-2}^6 f(x)dx = 467.28$

First, the indefinite integral of $f(x)$ is:

$\int f(x)dx = \int(4x + e^x)dx = \int 4xdx + \int e^x dx = 2x^2 + e^x + C = F(x)$

Applying the Fundamental Theorem of Calculus, we have:

$\int_{-2}^6 f(x)dx = F(6) - F(-2)$

$F(6) = 2(6)^2 + e^6 = 2(36) + 403.42 = 475.42$

$F(-2) = 2(-2)^2 + e^{-2} = 2(4) + 0.13 = 8.13$

$\int_{-2}^{6} f(x)dx = F(6) - F(-2) = 475.42 - 8.13 = 467.28$

8. Incorrect

If $f(x) = e^x + \sin(x)$ then, the $\int_{1}^{6} f(x)dx = 400.72$

First, the indefinite integral of $f(x)$ is:

$\int f(x)dx = \int[e^x + \sin(x)]dx = \int e^x dx + \int \sin(x)\, dx = e^x - \cos(x) + C = F(x)$

Applying the Fundamental Theorem of Calculus, we have:

$\int_{1}^{6} f(x)dx = F(6) - F(-2)$

$F(6) = e^6 - \cos(6) = 403.42 - 0.994 = 402.46$

$F(1) = e^1 - \cos(1) = 2.71 - 0.999 = 2.17$

$\int_{1}^{6} f(x)dx = F(6) - F(-2) = 402.43 - 1.71 = 400.29$

9. Incorrect

If $f(x) = 2e^x + 3x^2 - 4x + 5$ then, the $\int_{-3}^{8} f(x)dx = 6445.81$

First, the indefinite integral of $f(x)$ is:

$\int f(x)dx = \int(2e^x + 3x^2 - 4x + 5)dx = \int 2e^x dx + \int 3x^2 dx - \int 4x dx + \int 5dx = 2\,e^x + x^3 - 2x^2 + 5x + C = F(x)$

Applying the Fundamental Theorem of Calculus, we have:

$\int_{-3}^{8} f(x)dx = F(8) - F(-3)$

$F(8) = 2e^8 + (8)^3 - 2(8)^2 + 5(8) = 5961.91 + 512 - 128 + 40 = 6385.92$

$F(-3) = 2e^{-3} + (-3)^3 - 2(-3)^2 + 5(-3) = 0.099 - 27 - 18 - 15 = -59.9$

$\int_{-3}^{8} f(x)dx = F(8) - F(-3) = 6385.91 - (-59.9) = 6445.82$

10. Correct

If $f(x) = x^2 - 3x + 7$ then, the $\int_{-3}^{3} f(x)dx = 60$

First, the indefinite integral of $f(x)$ is:

$\int f(x)dx = \int(x^2 - 3x + 7)dx = \int x^2 dx - \int 3x dx + \int 7dx = \frac{1}{3}x^3 - \frac{3}{2}x^2 + 7x + C = F(x)$

Applying the Fundamental Theorem of Calculus, we have:

$\int_{-3}^{3} f(x)dx = F(3) - F(-3)$

$F(3) = \frac{1}{3}(3)^3 - \frac{3}{2}(3)^2 + 7(3) = 9 - 13.5 + 21 = 16.5$

$F(-3) = \frac{1}{3}(-3)^3 - \frac{3}{2}(-3)^2 + 7(-3) = -9 - 13.5 - 21 = -43.5$

$\int_{-3}^{3} f(x)dx = F(3) - F(-3) = 16.5 - (-43.5) = 60$

Chapter 4. E. a. Antiderivatives of functions

1. Correct

If $f(x) = \sin(x) + 2x - \frac{3x^2+3x}{x}$

then, the antiderivative of $f(x)$ will be: $F(x) = -\cos(x) - \frac{x^2}{2} - 3x + C$

$f(x) = \sin(x) + 2x - \frac{3x^2+3x}{x} = \sin(x) + 2x - 3x - 3 = \sin(x) - x - 3$

$F(x) = \int f(x)dx = \int(\sin(x) - x - 3)dx = \int[\sin(x)]dx - \int xdx - \int 3dx = -\cos(x) - \frac{x^2}{2} - 3x + C$

2. Incorrect

If $f(x) = 5x + \frac{4x^3+3x^2-x}{3}$ then,

the antiderivative of $f(x)$ will be: $F(x) = \frac{x^4}{3} + \frac{x^3}{3} + \frac{7x^2}{3} + C$

$f(x) = 5x + \frac{4x^3+3x^2-x}{3} = 5x + \frac{4x^3}{3} + x^2 - \frac{x}{3} = \frac{4x^3}{3} + x^2 + \frac{14x}{3}$

$F(x) = \int f(x)dx = \int(\frac{4x^3}{3} + x^2 + \frac{14x}{3})dx = \int\left(\frac{4x^3}{3}\right)dx + \int x^2dx + \int\left(\frac{14x}{3}\right)dx = \frac{4x^4}{3*4} + \frac{x^3}{3} + \frac{14x^2}{3*2} = \frac{x^4}{3} + \frac{x^3}{3} + \frac{7x^2}{3} + C$

3. Correct

If $f(x) = 7\tan(x) - sec^2(x)$ then,
the antiderivative of $f(x)$ will be: $F(x) = -7\ln|\cos(x)| - \tan(x) + C$
$F(x) = \int[f(x)]dx = \int[7\tan(x) - sec^2(x)]dx = \int[7\tan(x)]dx - \int[sec^2(x)]dx = -7\ln|\cos(x)| - \tan(x) + C$

4. Incorrect

If $f(x) = e^x - \frac{x^2-3x-10}{x+2} + 7$ then,

the antiderivative of $f(x)$ will be: $F(x) = e^x - \frac{x^2}{2} + 12x + C$

$f(x) = e^x - \frac{x^2-3x-10}{x+2} + 7 = e^x - \frac{(x+2)(x-5)}{x+2} + 7 = e^x - (x-5) + 7 = e^x - x + 12$

$F(x) = \int[f(x)]dx = \int(e^x - x + 12)dx = \int e^x dx - \int xdx + \int 12dx = e^x - \frac{x^2}{2} + 12x + C$

5. Correct

If $f(x) = \cos(x) - \frac{2x^2+3x}{x} + 3$ then,
the antiderivative of $f(x)$ will be: $F(x) = \sin(x) - x^2 + C$

$f(x) = \cos(x) - \frac{2x^2+3x}{x} + 3 = \cos(x) - 2x - 3 + 3 = \cos(x) - 2x$

$F(x) = \int[f(x)]dx = \int[\cos(x) - 2x]dx = \int[\cos(x)]dx - \int 2xdx = \sin(x) - x^2 + C$

6. Incorrect

If $f(x) = \frac{2x^2+3x}{x^2} - 5\sec^2(x)$ then,

the antiderivative of $f(x)$ will be: $F(x) = 2x + 3ln|x| - 5\tan(x) + C$

$f(x) = \frac{2x^2+3x}{x^2} - 5\sec^2(x) = \frac{2x^2}{x^2} + \frac{3x}{x^2} - 5\sec^2(x) = 2 + \frac{3}{x} - 5\sec^2(x)$

$F(x) = \int[f(x)]dx = \int[2 + \frac{3}{x} - 5\sec^2(x)]dx = \int 2dx + \int \frac{3}{x}dx - \int[5\sec^2(x)]dx = 2x + 3ln|x| - 5\tan(x) + C$

7. Correct

If $f(x) = \frac{sin^2(x)+cos^2(x)}{x} - \mathrm{e}^x$ then,

the antiderivative of $f(x)$ will be: $F(x) = ln|x| - \mathrm{e}^x + C$

$f(x) = \frac{sin^2(x)+cos^2(x)}{x} - \mathrm{e}^x = \frac{1}{x} - \mathrm{e}^x$

$F(x) = \int[f(x)]dx = \int[\frac{1}{x} - \mathrm{e}^x]dx = \int \frac{1}{x}dx - \int edx = ln|x| - \mathrm{e}^x + C$

8. Incorrect

If $f(x) = \frac{sin^2(x)-cos^2(x)}{\sin(x)+\cos(x)} + 1 + \ln(x)$ then,

the antiderivative of $f(x)$ will be: $F(x) = -\cos(x) - \sin(x) + xln(x) + C$

$f(x) = \frac{sin^2(x)-cos^2(x)}{\sin(x)+\cos(x)} + 1 + \ln(x) = \frac{[\sin(x)-\cos(x)][\sin(x)+\cos(x)]}{\sin(x)+\cos(x)} + 1 + \ln(x) = \sin(x) - \cos(x) + 1 + \ln(x)$

$F(x) = \int[f(x)]dx = \int[\sin(x) - \cos(x) + 1 + \ln(x)]dx = \int[\sin(x)]\,dx - \int[\cos(x)]\,dx + \int dx + \int \ln(x)\,dx = -\cos(x) - \sin(x) + x + xln(x) - x = -\cos(x) - \sin(x) + xln(x)$

9. Correct

If $f(x) = sin^2(x) + cos^2(x) + 1 - \ln(x)$ then,

the antiderivative of $f(x)$ will be: $F(x) = 3x - xln(x) + C$

$f(x) = sin^2(x) + cos^2(x) + 1 - \ln(x) = 1 + 1 - \ln(x) = 2 - \ln(x)$

$F(x) = \int[f(x)]dx = \int[2 - \ln(x)]dx = \int 2dx - \int \ln(x)\,dx = 2x - xln(x) + x + C = 3x - xln(x) + C$

10. Incorrect

If $f(x) = \frac{2}{\sqrt{x}} - \sqrt{x}$ then,

the antiderivative of $f(x)$ will be: $F(x) = 4\sqrt{x} - \frac{3}{2}x\sqrt{x} + C$

$f(x) = 2x^{-\frac{1}{2}} - x^{\frac{1}{2}}$

$F(x) = \int[f(x)]dx = \int[2x^{-\frac{1}{2}} - x^{\frac{1}{2}}]dx = \int(2x^{-\frac{1}{2}})dx - \int(x^{\frac{1}{2}})dx = 2\frac{x^{\frac{1}{2}}}{\frac{1}{2}} - \frac{2}{3}x^{\frac{3}{2}} + C = 4\sqrt{x} - \frac{2}{3}x\sqrt{x} + C$

Chapter 4. E. b. Methods of Integration - Substitution

1. Correct

If $f(x) = 4(2x + 1)$ then, $F(x) = \int f(x)dx = (2x + 1)^2 + C$
$u = 2x + 1$
$u' = \frac{du}{dx} = 2\ so, dx = \frac{du}{2}$
$\int 4u\frac{du}{2} = \int 2udu = 2\frac{u^2}{2} = u^2$
But $u = 2x + 1$ so $F(x) = \int f(x)dx = (2x + 1)^2 + C$

2. Incorrect

If $f(x) = 30x(5x^2 - 3)^2$ then, $F(x) = \int f(x)dx = (5x - 3)^3 + C$
$u = 5x^2 - 3$
$u' = \frac{du}{dx} = 10x\ so, dx = \frac{du}{10x}$
$\int 30xu^2\frac{du}{10x} = \int 3u^2du = 3\frac{u^3}{3} = u^3$
But $u = 5x^2 - 3$ so $F(x) = \int f(x)dx = (5x^2 - 3)^3 + C$

3. Correct

If $f(x) = 2xcos(x^2)$ then, $F(x) = \int f(x)dx = \sin(x^2) + C$
$u = x^2$
$u' = \frac{du}{dx} = 2x\ so, dx = \frac{du}{2x}$
$\int 2xcos(u)\frac{du}{2x} = \int cos\ (u)du = \sin\ (u) = \sin(x^2) + C$

4. Incorrect

If $f(x) = \frac{5}{3x-73}$ then, $F(x) = \int f(x)dx = \frac{5}{3}ln|3x - 73| + C$
$u = 3x - 73$
$u' = \frac{du}{dx} = 3\ so, dx = \frac{du}{3}$
$\int \frac{5}{u}\frac{du}{3} = \frac{5}{3}ln|u| + C$
$F(x) = \int f(x)dx = \frac{5}{3}ln|3x - 73| + C$

5. Incorrect

If $f(x) = \frac{7\ln(x)}{5x}$ then, $F(x) = \int f(x)dx = \frac{7}{10}[ln|x|]^2 + C$
$u = \ln(x)$
$u' = \frac{du}{dx} = \frac{1}{x}\ so, dx = xdu$

$\int f(x)\,dx = \int \frac{7u}{5x} x du = \int \frac{7u}{5} du = \frac{7u^2}{5*2} = \frac{7}{10}u^2 + C$

$F(x) = \int f(x)dx = \frac{7}{10}[ln|x|]^2 + C$

6. Correct

If $f(x) = 3x^3e^{x^4}$then, $F(x) = \int f(x)dx = \frac{3}{4}e^{x^4} + C$

$u = x^4\ so, u' = \frac{du}{dx} = 4x^3$

$dx = \frac{du}{4x^3}$

$\int 3x^3e^u \frac{du}{4x^3} = \int \frac{3}{4}e^u du = \frac{3}{4}e^u + C$

$F(x) = \int f(x)dx = \frac{3}{4}e^{x^4} + C$

7. Incorrect

If $f(x) = (6x-4)(3x^2-4x+5)^2$then, $F(x) = \int f(x)dx = \frac{1}{3}(3x^2-4x+5)^3 + C$

$u = 3x^2 - 4x + 5\ so, u' = \frac{du}{dx} = 6x - 4$

$dx = \frac{du}{6x-4}$

$\int (6x-4)u^2 \frac{du}{6x-4} = \int u^2 du = \frac{1}{3}u^3 + C$

$F(x) = \int f(x)dx = \frac{1}{3}(3x^2-4x+5)^3 + C$

8. Correct

If $f(x) = \frac{8x^3+3}{\sqrt{2x^4+3x-5}}$ then, $F(x) = \int f(x)dx = 2\sqrt{2x^4+3x-5} + C$

$u = 2x^4 + 3x - 5\ so, u' = \frac{du}{dx} = 8x^3 + 3$

$dx = \frac{du}{8x^3+3}$

$\int \frac{(8x^3+3)du}{(8x^3+3)\sqrt{u}} = \int \frac{du}{\sqrt{u}} = \int u^{-\frac{1}{2}}du = \frac{u^{-\frac{1}{2}+1}}{\frac{1}{2}} = 2\sqrt{u} + C$

$F(x) = \int f(x)dx = 2\sqrt{2x^4+3x-5} + C$

9. Incorrect

If $f(x) = \frac{9x^2-8x}{\sqrt{3x^3-4x^2+5}}$ then, $\int_1^3 f(x) = F(3) - F(1) = 10.14$

$u = 3x^3 - 4x^2 + 5\ so, u' = \frac{du}{dx} = 9x^2 - 8x$

$dx = \frac{du}{9x^2-8x}$

$\int \frac{(9x^2-8x)du}{(9x^2-8x)\sqrt{u}} = \int \frac{du}{\sqrt{u}} = 2\sqrt{u} + C$

$\int_1^3 f(x) = F(3) - F(1) = 2\sqrt{3(3)^3 - 4(3)^2 + 5} - 2\sqrt{3(1)^3 - 4(1)^2 + 5} = 2\sqrt{81 - 36 + 5} - 2\sqrt{3 - 4 + 5} = 2\sqrt{50} - 2\sqrt{4} = 14.14 - 4 = 10.14$

10. Incorrect

If $f(x) = \frac{1}{\sqrt{x^2+9}} = \frac{1}{\sqrt{x^2+3^2}}$ then, $F(x) = \int f(x)dx = \ln\left|\sec[\tan^{-1}(\frac{x}{3})] + \tan[\tan^{-1}\left(\frac{x}{3}\right)]\right| + C$

$x = 3\tan(\Theta)\ so, x' = \frac{dx}{d\Theta} = 3sec^2(\Theta)\ ;\ \tan(\Theta) = \frac{x}{3}\ so, \Theta = \tan^{-1}(\frac{x}{3})$

$dx = 3sec^2(\Theta)\ d\Theta$

$tan^2(\Theta) + 1 = \frac{sin^2(\Theta)}{cos^2(\Theta)} + 1 = \frac{sin^2(\Theta)+cos^2(\Theta)}{cos^2(\Theta)} = \frac{1}{cos^2(\Theta)} = sec^2(\Theta)$

$\int \frac{3sec^2(\Theta)\ d\Theta}{\sqrt{9tan^2(\Theta)+9}} = \int \frac{3sec^2(\Theta)\ d\Theta}{\sqrt{9[tan^2(\Theta)+1]}} = \int \frac{3sec^2(\Theta)\ d\Theta}{3\sqrt{[tan^2(\Theta)+1]}} = \int \frac{sec^2(\Theta)\ d\Theta}{\sqrt{sec^2(\Theta)}} = \int \frac{sec^2(\Theta)\ d\Theta}{\sec(\Theta)} = \int \sec(\Theta)\ d\Theta = \ln|\sec(\Theta) + \tan(\Theta)| + C$

$F(x) = \int f(x)dx = \ln\left|\sec[\tan^{-1}(\frac{x}{3})] + \tan[\tan^{-1}\left(\frac{x}{3}\right)]\right| + C$

Chapter 4. E. c. Methods of Integration – By parts

1. Correct

If $f(x) = x^4\ln(x)$ then $F(x) = \int f(x)dx = \frac{1}{5}x^5\left[\ln(x) - \frac{1}{5}\right] + C$

$u = \ln(x)\ so\ u' = \frac{du}{dx} = \frac{1}{x}$

$du = \frac{1}{x}dx$

$dv = x^4dx\ so\ v = \int x^4dx = \frac{1}{5}x^5 + C$

By using the integration by parts formula, we have:

$\int udv = uv - \int vdu$

$\int \ln(x)(x^4dx) = \ln(x)\frac{1}{5}x^5 - \int \frac{1}{5}x^5\left(\frac{1}{x}\right)dx = \ln(x)\frac{1}{5}x^5 - \frac{1}{5}\int x^4dx = \ln(x)\frac{1}{5}x^5 - \frac{1}{5} * \frac{x^5}{5} = \ln(x)\frac{1}{5}x^5 - \frac{x^5}{25} = \frac{1}{5}x^5\left[\ln(x) - \frac{1}{5}\right] + C$

2. Correct

If $f(x) = x^3cos(x)$ then $F(x) = \int f(x)dx = \sin(x)(x^3) + 3x^2\cos(x) + 6xsin(x) + 6\cos(x) + C$

$u = x^3 so\ u' = \frac{du}{dx} = 3x^2$

$du = 3x^2dx$

$dv = \cos(x)dx\ so\ v = \int \cos(x)dx = \sin(x) + C$

By using the integration by parts formula, we have:

$\int udv = uv - \int vdu$

$\int \cos(x)(x^3dx) = \sin(x)(x^3) - 3\int x^2[\sin(x)\,dx] =$

$3\int x^2[\sin(x)\,dx] =$

$u = x^2, u' = \frac{du}{dx} = 2x\ so, du = 2xdx$

$dv = \sin(x)\,dx, so\ v = \int \sin(x)\,dx = -\cos(x) + C$

$3\int x^2[\sin(x)\,dx] = 3\{[x^2(-\cos x)] - \int[-\cos(x)]2xdx\} = -3x^2\cos(x) + 6\int xcos(x)dx$

So,

$\int \cos(x)\,(x^3dx) = \sin(x)\,(x^3) - [-3x^2\cos(x) + 6\int xcos(x)dx]$

$\int x\,cos(x)dx =$

$u = x,, u' = \frac{du}{dx} = 1\ so, du = dx$

$dv = \cos(x), v = \int \cos(x)\,dx = \sin(x)$

$\int udv = uv - \int vdu$

$\int x\,cos(x)dx = xsin(x) - \int \sin(x)\,dx = xsin(x) - [-\cos(x)] = xsin(x) + \cos(x)$

$\int \cos(x)\,(x^3dx) = \sin(x)\,(x^3) - [-3x^2\cos(x) + 6\int xcos(x)dx] = \sin(x)\,(x^3) + 3x^2\cos(x) +$
$6[xsin(x) + \cos(x)] = \sin(x)\,(x^3) + 3x^2\cos(x) + 6xsin(x) + 6\cos(x)$

3. Incorrect

If $f(x) = \frac{\ln(x)}{x^3}$ then $F(x) = \int f(x)dx = -\frac{1}{2x^2}\left[\ln(x) + \frac{1}{2}\right] + C$

$u = \ln(x)\ so\ u' = \frac{du}{dx} = \frac{1}{x}$

$du = \frac{1}{x}dx$

$dv = \frac{1}{x^3}dx\ so\ v = \int \frac{1}{x^3}dx = \int x^{-3}dx = -\frac{1}{2}x^{-2} + C$

By using the integration by parts formula, we have:

$\int udv = uv - \int vdu$

$\int \ln(x)\left(\frac{1}{x^3}dx\right) = \ln(x)\left(-\frac{1}{2}x^{-2}\right) - \int\left(-\frac{1}{2}x^{-2}\right)\left(\frac{1}{x}dx\right) = -\frac{\ln(x)}{2x^2} + \int\left(\frac{1}{2}x^{-3}dx\right) = -\frac{\ln(x)}{2x^2} - \frac{1}{4x^2} =$
$-\frac{1}{2x^2}\left[\ln(x) + \frac{1}{2}\right] + C$

4. Incorrect

If $f(x) = x^2sin(x)$ then $F(x) = \int f(x)dx = -x^2[\cos(x)] + 2x\sin(x) + 2\cos(x) + C$

$u = x^2\ so\ u' = \frac{du}{dx} = 2x$

$du = 2xdx$

$dv = \sin(x)\,dx\ so\ v = \int \sin(x)\,dx = -\cos(x) + C$

By using the integration by parts formula, we have:

$\int udv = uv - \int vdu$

$\int x^2\sin(x)\,dx = x^2[-\cos(x)] - \int[-\cos(x)]2xdx = -x^2[\cos(x)] + 2\int \cos(x)\,xdx + C$

We use the integration by parts formula again for $\int \cos(x)xdx$;

$u = x\ so, u' = \frac{du}{dx} = 1$

$du = dx$

$dv = \cos(x)\,dx\ so, v = \int \cos(x)\,dx = \sin(x) + C$

$\int udv = uv - \int vdu$

$\int xcos(x)dx = xsin(x) - \int \sin(x)\,dx = xsin(x) + \cos(x) + C$

So, we have

$\int x^2 \sin(x)\,dx = x^2[-\cos(x)] - \int[-\cos(x)]2xdx = -x^2[\cos(x)] + 2\int \cos(x)xdx = -x^2[\cos(x)] + 2[xsin(x) + \cos(x)] = -x^2[\cos(x)] + 2x\sin(x) + 2\cos(x) + C$

5. Correct

If $f(x) = e^x cos(x)$ then $F(x) = \int f(x)dx = \frac{1}{2}e^x[\sin(x) + \cos(x)] + C$

$u = e^x\ so\ u' = \frac{du}{dx} = e^x$

$du = e^x dx$

$dv = \cos(x)\,dx\ so\ v = \int \cos(x)\,dx = \sin(x) + C$

By using the integration by parts formula, we have:

$\int udv = uv - \int vdu$

**$\int e^x \cos(x)\,dx = e^x \sin(x) - \int e^x \sin(x)\,dx + C$

We use the integration by parts formula again for $\int e^x \sin(x)\,dx$:

$u = e^x\ so\ u' = \frac{du}{dx} = e^x$

$du = e^x dx$

$dv = \sin(x)\,dx\ so\ v = \int \sin(x)\,dx = -\cos(x) + C$

By using the integration by parts formula, we have:

$\int udv = uv - \int vdu$

$\int e^x \sin(x)\,dx = e^x[-\cos(x)] - \int[-\cos(x)]e^x dx = -e^x[\cos(x) + \int[\cos(x)]e^x dx + C$

So, the original integral in formula **

$\int[\cos(x)]e^x = e^x \sin(x) - \{-e^x[\cos(x)] + \int[\cos(x)]e^x dx\}$

$\int[\cos(x)]e^x = e^x \sin(x) + e^x \cos(x) - \int[\cos(x)]e^x dx$

$2\int[\cos(x)]e^x = e^x \sin(x) + e^x \cos(x) + C$

$\int[\cos(x)]e^x = \frac{1}{2}e^x[\sin(x) + \cos(x)] + C$

6. Incorrect

If $f(x) = x^2 e^x$ then $F(x) = \int f(x)dx = x^2 e^x - xe^x + e^x + C$

$u = x^2\ so\ u' = \frac{du}{dx} = 2x$

$du = 2xdx$

$dv = e^x dx\ so\ v = \int e^x dx = e^x + C$

By using the integration by parts formula, we have:

$\int udv = uv - \int vdu$

$\int x^2e^xdx = x^2e^x - \int e^x2xdx = x^2e^x - 2\int e^xxdx$

By using the integration by parts formula again for $\int e^xxdx$, we have:

$u = x\ so, so\ u' = \frac{du}{dx} = 1$

$du = dx$

$dv = e^xdx\ so\ v = \int e^xdx = e^x + C$

$\int udv = uv - \int vdu$

$\int e^xxdx = xe^x - \int e^xdx = xe^x - e^x + C$

So:

$\int x^2e^xdx = x^2e^x - 2(xe^x - e^x) = x^2e^x - 2xe^x + 2e^x + C$

7. Correct

If $f(x) = \frac{\ln(x)}{x}$ then $F(x) = \int f(x)dx = \frac{1}{2}ln^2(x) + C$

$u = \ln(x)\ so, so\ u' = \frac{du}{dx} = \frac{1}{x}$

$du = \frac{1}{x}dx$

$dv = \frac{1}{x}\ dx\ so\ v = \int \frac{1}{x}dx = \ln(x) + C$

$\int udv = uv - \int vdu$

$\int \frac{\ln(x)}{x}dx = \ln(x)\,[\ln(x)] - \int \frac{\ln(x)}{x}dx$

$2\int \frac{\ln(x)}{x}dx = ln^2(x) + C$

$\int \frac{\ln(x)}{x}dx = \frac{1}{2}ln^2(x) + C$

8. Incorrect

If $f(x) = x\sqrt{1 - sin^2(x)}$ then $F(x) = \int f(x)dx = xsin(x) + \cos(x) + C$

$f(x) = x\sqrt{1 - sin^2(x)} = x\sqrt{cos^2(x)} = xcos(x)$

$\int f(x)dx = \int xcos(x)dx$

$u = x\ so, u' = \frac{du}{dx} = 1$

$du = dx$

$dv = \cos(x)dx, so\ v = \int \cos(x)\,dx = \sin(x) + C$

$\int udv = uv - \int vdu$

$\int xcos(x)dx = xsin(x) - \int \sin(x)\,dx = xsin(x) - [-\cos(x)] = xsin(x) + \cos(x) + C$

9. Correct

If $f(x) = xln(x)$ then $F(x) = \int f(x)dx = \frac{x^2}{2}[\ln(x) - \frac{1}{2}] + C$

$u = ln(x)\ so, u' = \frac{du}{dx} = \frac{1}{x}$

$du = \frac{1}{x}dx$

$dv = xdx\ so, v = \int xdx = \frac{x^2}{2} + C$

$\int udv = uv - \int vdu$

$\int xln(x) = \ln(x)\frac{x^2}{2} - \int \frac{x^2}{2} * \frac{1}{x}dx = \ln(x)\frac{x^2}{2} - \int \frac{x}{2}dx = \ln(x)\frac{x^2}{2} - \frac{x^2}{4} = \frac{x^2}{2}\left[\ln(x) - \frac{1}{2}\right] + C$

10. Correct

If $f(x) = e^x \sin(2x)$ then $F(x) = \int f(x)dx = \frac{1}{5}sin(2x)e^x - \frac{2}{5}e^x cos(2x) + C$

$u = sin(2x)\ so, u' = \frac{du}{dx} = 2\cos(2x)$

$du = 2\cos(2x)dx$

$dv = e^x dx\ so, v = \int e^x dx = e^x + C$

$\int udv = uv - \int vdu$

$\int \sin(2x)\, e^x dx = sin(2x)e^x - \int e^x 2\cos(2x)\, dx = sin(2x)e^x - 2\int e^x \cos(2x)\, dx$

By using the integration by parts formula again for $\int e^x \cos(2x)dx$, we have:

$u = cos(2x)\ so, u' = \frac{du}{dx} = -2\sin(2x)$

$du = -2\sin(2x)dx$

$dv = e^x dx\ so, v = \int e^x dx = e^x + C$

$\int udv = uv - \int vdu$

$\int e^x \cos(2x)\, dx = e^x cos(2x) - \int e^x[-2\sin(2x)dx] = e^x cos(2x) + \int e^x[2\sin(2x)dx] = e^x cos(2x) + 2\int e^x \sin(2x)dx$

$\int \sin(2x)\, e^x dx = sin(2x)e^x - 2[e^x cos(2x) + 2\int e^x \sin(2x)\, dx] = sin(2x)e^x - 2e^x cos(2x) - 4\int e^x \sin(2x)\, dx$

$5\int e^x \sin(2x)\, dx = sin(2x)e^x - 2e^x cos(2x) + C$

$\int e^x \sin(2x)\, dx = \frac{1}{5}sin(2x)e^x - \frac{2}{5}e^x cos(2x) + C$

Chapter 4. F. a. Aria under a curve, volume of solids, average value of functions

PAGE 1

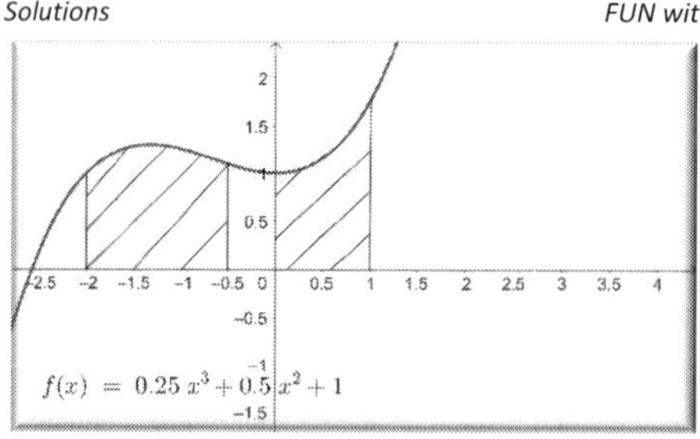

1. Incorrect

The area under $f(x) = \frac{1}{4}x^3 + \frac{1}{2}x^2 + 1$ between x=-2 and x=-0.5 is: 1.813

$Area = \int_{-2}^{-0.5} f(x)dx = F(-0.5) - F(-2)$

$F(x) = \int f(x)dx = \int \left[\frac{1}{4}x^3 + \frac{1}{2}x^2 + 1\right] dx =$

$\int \frac{1}{4}x^3 dx + \int \frac{1}{2}x^2 dx + \int dx = \frac{1}{16}x^4 + \frac{1}{6}x^3 + x + C$

$F(-0.5) = \frac{1}{16}(-0.5)^4 + \frac{1}{6}(-0.5)^3 + (-0.5) =$

$0.003 - 0.02 - 0.5 = -0.517$

$F(-2) = \frac{1}{16}(-2)^4 + \frac{1}{6}(-2)^3 + (-2) = 1 - 1.33 - 2 = -2.33$

$Area = \int_{-2}^{-0.5} f(x)dx = F(-0.5) - F(-2) = -0.517 + 2.33 = 1.813$

2. Correct

The area under $f(x) = \frac{1}{4}x^3 + \frac{1}{2}x^2 + 1$ between x=0 and x=1 is: 1.223

$Area = \int_0^1 f(x)dx = F(1) - F(0)$

$F(x) = \int f(x)dx = \int \left[\frac{1}{4}x^3 + \frac{1}{2}x^2 + 1\right] dx = \int \frac{1}{4}x^3 dx + \int \frac{1}{2}x^2 dx + \int dx = \frac{1}{16}x^4 + \frac{1}{6}x^3 + x + C$

$F(1) = \frac{1}{16}(1)^4 + \frac{1}{6}(1)^3 + (1) = 0.0625 + 0.16 + 1 = 1.223$

$F(0) = \frac{1}{16}(0)^4 + \frac{1}{6}(0)^3 + (0) = 0$

$Area = \int_0^1 f(x)dx = F(1) - F(0) = 1.223$

3. Incorrect

The area under $f(x) = 0.18x^4 - 0.6x^3 + 0.5x^2 + x + 2$ between x=0.1 and x=3 is:11.39

$Area = \int_{0.1}^{3} f(x)dx = F(3) - F(0.1)$

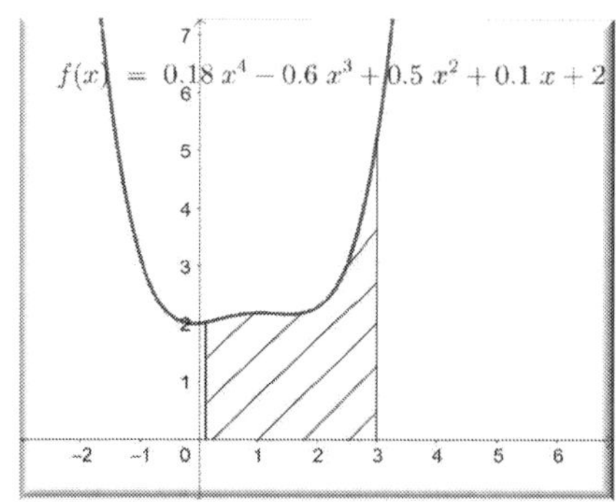

$F(x) = \int f(x) = \int [0.18x^4 - 0.6x^3 + 0.5x^2 + x + 2]dx = \int 0.18x^4 dx - \int 0.6x^3 dx + \int 0.5x^2 + \int x dx + \int 2dx = \frac{0.18}{5}x^5 - \frac{0.6}{4}x^4 + \frac{0.5}{3}x^3 + \frac{1}{2}x^2 + 2x + C$

$F(3) = \frac{0.18}{5}(3)^5 - \frac{0.6}{4}(3)^4 + \frac{0.5}{3}(3)^3 + \frac{1}{2}(3)^2 + 2(3) = 8.74 - 12.15 + 4.5 + 4.5 + 6 = 11.59$

$F(0.1) = \frac{0.18}{5}(0.1)^5 - \frac{0.6}{4}(0.1)^4 + \frac{0.5}{3}(0.1)^3 + \frac{1}{2}(0.1)^2 + 2(0.1) = 0 - 0.00001 + 0.0001 + 0.005 + 0.2 = 0.2$

$Area = \int_{0.1}^{3} f(x)dx = F(3) - F(0.1) = 11.59 - 0.2 = 11.39$

4. Incorrect

The volume of the object that is made by the function $f(x) = x^2 + 2$ that rotate around x axis between x=0 and x=2 is: V=25.06 π

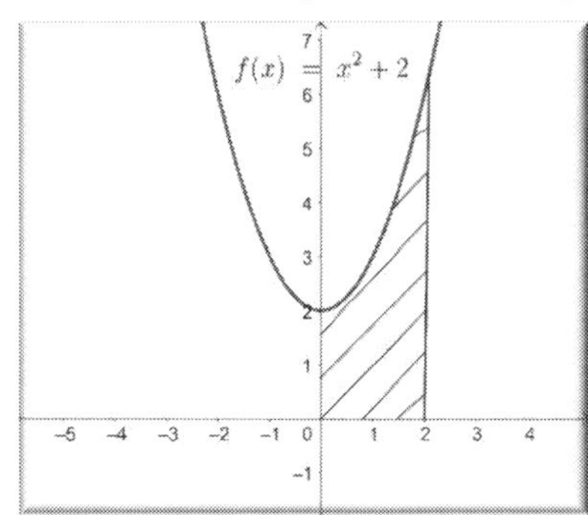

$V = \int_0^2 \pi[f(x)]^2 dx = \pi \int_0^2 (x^2 + 2)^2 dx = \pi \int_0^2 (x^4 + 4x^2 + 4) dx = \pi[F(2) - F(0)]$

$F(x) = \int (x^4 + 4x^2 + 4) dx = \int x^4 dx + \int 4x^2 dx + \int 4dx = \frac{1}{5}x^5 + \frac{4}{3}x^3 + 4x + C$

$F(2) = \frac{1}{5}(2)^5 + \frac{4}{3}(2)^3 + 4(2) = 6.4 + 10.6 + 8 = 25.06$

$F(0) = \frac{1}{5}(0)^5 + \frac{4}{3}(0)^3 + 4(0) = 0$

$V = \pi[F(2) - F(0)] = 25.06\pi$

5. Correct

The volume of the object that is made by the function $f(x) = -0.3x^2 + 4$ that rotate around x axis between x=0.5 and x=3 is: V= 22.87π

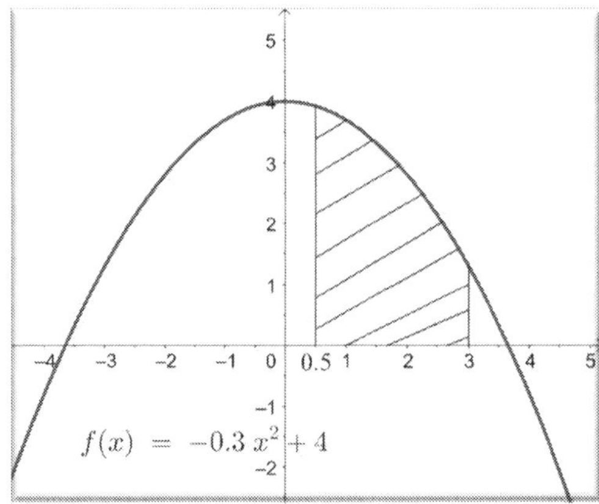

$V = \int_{0.5}^{3} \pi[f(x)]^2 dx = \pi \int_{0.5}^{3} (-0.3x^2 + 4)^2 dx = \pi \int_{0.5}^{3} (0.09x^4 - 2.4x^2 + 16) dx = \pi[F(3) - F(0.5)]$

$F(x) = \int (0.09x^4 - 2.4x^2 + 16) dx = \int 0.09x^4 dx - \int 2.4x^2 dx + \int 16 dx = \frac{0.09}{5}x^5 - \frac{2.4}{3}x^3 + 16x + C$

$F(3) = \frac{0.09}{5}(3)^5 - \frac{2.4}{3}(3)^3 + 16(3) = 4.37 - 21.6 + 48 = 30.77$

$F(0.5) = \frac{0.09}{5}(0.5)^5 - \frac{2.4}{3}(0.5)^3 + 16(0.5) = 0.00056 - 0.1 + 8 = 7.9$

$V = \pi[F(3) - F(0.5)] = \pi(30.77 - 7.9) = 22.87\pi$

6. Incorrect

The volume of the sphere that is made by the function $4 = x^2 + y^2$ that rotate around x axis between x=-2 and x=2 is: V= 10.66π

We have:

$x^2 = 4 - y^2$

Area of the circle with the radius x is:

$A = \pi x^2 = \pi(4 - y^2)$

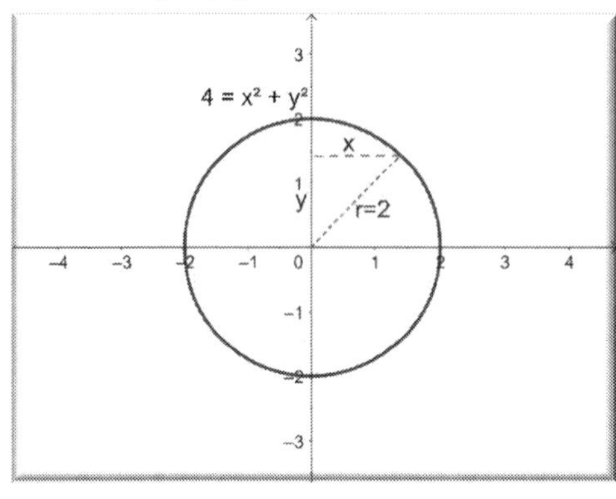

The volume is:

$V(y) = \int A(y)dy = \int \pi(4 - y^2)dy = 4\pi \int dy - \pi \int y^2 dy = 4\pi y - \frac{\pi}{3}y^3 + C$

If we consider the limits as -2 and 2 we will have:

$V = \int_{-2}^{2}(\pi(4 - y^2)dy = F(2) - F(-2)$

$F(2) = 4\pi(2) - \frac{\pi}{3}(2)^3 = 8\pi - 2.66\pi = 5.33\pi$

$F(-2) = 4\pi(-2) - \frac{\pi}{3}(-2)^3 = -8\pi + 2.66\pi = -5.33\pi$

$V = \int_{-2}^{2}(\pi(4 - y^2)dy = F(2) - F(-2) = 5.33\pi - (-5.33)\pi = 10.66\pi$

7. Correct

The volume of the object that is made by the function $f(x) = x^2$ that rotate around x axis between x=0 and x=3 is: 48.6π

$V = \int_0^3 \pi[f(x)]^2 dx = \pi \int_0^3 (x^2)^2 dx = \pi \int_0^3 (x^4)dx = \pi[F(3) - F(0)]$

$F(x) = \int x^4 dx = \frac{1}{5}x^5 + C$

$F(3) = \frac{1}{5}(3)^5 = 48.6$

$F(0) = \frac{1}{5}(0)^5 = 0$

$V = \pi[F(3) - F(0)] = \pi(48.6 - 0) = 48.6\pi$

8. Incorrect

The volume of the object that is made by the function $f(x) = x$ that rotate around x axis between x=1 and x=5 is: 41.33 π

$V = \int_1^5 \pi[f(x)]^2 dx = \pi \int_1^5 (x)^2 dx = \pi \int_1^5 (x^2)dx = \pi[F(5) - F(1)]$

$F(x) = \int x^2 dx = \frac{1}{3}x^3 + C$

$F(5) = \frac{1}{3}(5)^3 = 41.66$

$F(1) = \frac{1}{3}(1)^3 = 0.33$

$V = \pi[F(5) - F(1)] = \pi(41.66 - 0.33) = 41.33\pi$

9. Correct

The average value of the function $f(x) = 2x + \sin(x)$ on the interval [1,3] is: 4

$\int_1^3 f(x)dx = F(3) - F(1)$

$F(x) = \int f(x)dx = \int[2x + \sin(x)]\, dx = \int 2xdx + \int \sin(x)\, dx = x^2 - \cos(x) + C$

$F(3) = (3)^2 - \cos(3) = 9 - 0.998 = 8.002$

$F(1) = (1)^2 - \cos(1) = 1 - 0.999 = 0$

$\int_1^3 f(x)dx = F(3) - F(1) = 8 - 0 = 8$

$Average\ value = \frac{1}{3-1}\int_1^3 f(x)dx = \frac{8}{2} = 4$

10. Incorrect

The average value of the function $f(x) = x^2 + \frac{1}{x}$ on the interval [1,5] is: 10.73

$\int_1^5 f(x)dx = F(5) - F(1)$

$F(x) = \int f(x)dx = \int(x^2 + \frac{1}{x})dx = \int x^2dx + \int \frac{1}{x}dx = \frac{1}{3}x^3 + \ln(x) + C$

$F(5) = \frac{1}{3}(5)^3 + \ln(5) = 41.66 + 1.6 = 43.26$

$F(1) = \frac{1}{3}(1)^3 + \ln(1) = 0.33 + 0 = 0.33$

$\int_1^5 f(x)dx = F(5) - F(1) = 43.26 - 0.33 = 42.93$

$Average\ value = \frac{1}{5-1}\int_1^5 f(x)dx = \frac{42.93}{4} = 10.73$

Chapter 4. F. b. Differential equations, Initial value problems, Slope fields

The trajectory of the Space Shuttle in the first minutes is represented by:
$h(t) = 2008 - 0.047t^3 + 18.3t^2 - 345t$

1. Correct

The velocity of the Space Shuttle at 20 seconds is: 330.6 m/s.
$h'(t) = -(0.047t^3)' + (18.3t^2)' - (345t)' = -0.047(3)t^2 + 18.3(2)t - 345$
$h'(20) = -0.047(3)(20)^2 + 18.3(2)(20) - 345 = -56.4 + 732 - 345 = 330.6m/s$

2. Incorrect

The acceleration of the Space Shuttle at 20 seconds is $30.9\ m/s^2$.
$h''(t) = (-0.141t^2)' + (36.6t)' = -0.141(2)t + 36.6$
$h''(20) = -0.282(20) + 36.6 = -5.64 + 36.6 = 30.9\ m/s^2$
A formula one speed car has the acceleration formula: a(t)=t+4 m/s^2.

3. Correct

The velocity of the car after 10 seconds is: $90\ m/s$

$v(t) = \int a(t)dt = \int (\mathrm{t} + 4)\mathrm{dt} = \frac{1}{2}t^2 + 4t + C$

$v(10) = \frac{1}{2}(10)^2 + 4(10) = 50 + 40 = 90\ m/s.$

4. Incorrect

The distance traveled by the car after 10 seconds is: $366.6\ m$.

$s(t) = \int v(t)dt = \int (\frac{1}{2}t^2 + 4t)dt = \frac{1}{6}t^3 + \frac{4}{2}t^2 + C$

$s(10) = \frac{1}{6}(10)^3 + 2(10)^2 = 166.6 + 200 = 366.6\ m.$

5. Correct

If the differential equation is $y' = \frac{dy}{dx} = 15 + x$, the initial condition is y(0)=3, then:

$y = 15x + \frac{1}{2}x^2 + 3$

$dy = (15 + x)dx$

We integrate both sides;

$\int dy = \int (15 + x)dx = \int 15dx + \int xdx = 15x + \frac{1}{2}x^2 + C$

$y = 15x + \frac{1}{2}x^2 + C$

We have that y (0) =3

$3 = 15(0) + \frac{1}{2}(0)^2 + C$

$3 = C$

$y = 15x + \frac{1}{2}x^2 + 3$

6. Incorrect

If the differential equation is $y' = \frac{dy}{dx} = 2x^2 - 3x$, the initial condition is y(1)=2, then:

$y = \frac{2}{3}x^3 - \frac{3}{2}x^2 + 2.83$

$dy = (2x^2 - 3x)dx$

We integrate both sides:

$\int dy = \int (2x^2 - 3x)dx = \int 2x^2dx - \int 3xdx = \frac{2}{3}x^3 - \frac{3}{2}x^2 + C$

$y = \frac{2}{3}x^3 - \frac{3}{2}x^2 + C$ so, $2 = \frac{2}{3}(1)^3 - \frac{3}{2}(1)^2 + C$

$2 - \frac{2}{3} + \frac{3}{2} = C$

$\frac{12}{6} - \frac{4}{6} + \frac{9}{6} = \frac{17}{6} = 2.83 = C$

$y = \frac{2}{3}x^3 - \frac{3}{2}x^2 + 2.83$

7. Incorrect

If the differential equation is $y' = \frac{dy}{dx} = e^x + 4x$, the initial condition is y(0)=5, then:

$y = e^x + 2x^2 + 4$

$dy = (e^x + 4x)dx$

We integrate both sides:

$\int dy = \int (e^x + 4x)dx = \int e^x dx + \int 4x dx = e^x + 2x^2 + C$

$y = e^x + 2x^2 + C$

$5 = e^0 + 2(0)^2 + C$

$5 = 1 + C$

$4 = C$

$y = e^x + 2x^2 + 4$

8. Correct

One of the functions whose derivative is $f(x) = 4x$ is $F(x) = 2x^2 + 37$

$F(x) = \int 4x dx = 2x^2 + C$

So, one of the functions is $F(x) = 2x^2 + 37$

9. Incorrect

One of the functions whose derivative is $f(x) = 3x^{-1}$ is $F(x) = 3\ln(x) + 3$

$F(x) = \int 3x^{-1} dx = 3\ln(x) + C$

So, one of the functions is $F(x) = 3\ln(x) + 3$

10. Incorrect

One of the functions whose derivative is $f(x) = e^x$ is $F(x) = e^x + 3$

$F(x) = \int e^x dx = e^x + C$

So, one of the functions is $F(x) = e^x + 3$

PICTURES USED

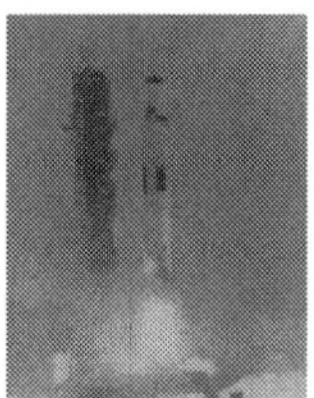

Apollo 10 lift-off

https://images-assets.nasa.gov/image/S69-34143/S69-34143~small.jpg

Fallingwater Drawing– Frank Lloyd Wright

Photo Marcel Sincraian

Athabasca Glacier – Alberta, Canada

Photo Marcel Sincraian

Mount Andromeda left of Athabasca Glacier – Alberta, Canada

Photo Marcel Sincraian

Grizzly bear – Jasper National Park, Alberta

Photo – Marcel Sincraian

Sagrada Famillia -Barcelona, Spain

Photo – Marcel Sincraian

Glass Pyramid – Louvre Museum, Paris, France

Photo – Marcel Sincraian

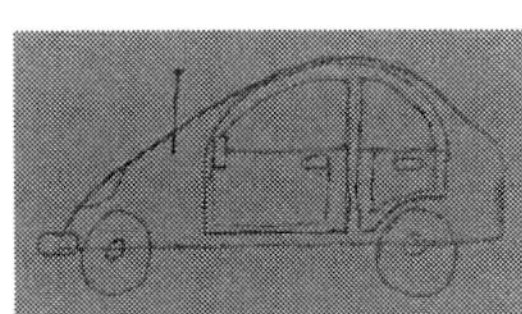

Car drawing

Photo Marcel Sincraian

Horse drawing

Photo Marcel Sincraian

Elk – Banff National Park

Photo Marcel Sincraian

FORMULAS

Differentiation Formulas

$x' = 0$

$[f(x) * g(x)]' = f'(x) * g(x) + f(x) * g'(x)$

$[f(x) \mp g(x)]' = f'(x) \mp g'(x)$

$[c * f(x)]' = c * f'(x)$

$\left[\frac{f(x)}{g(x)}\right]' = \frac{[f(x)]' g(x) - f(x)[g(x)]'}{[g(x)]^2}$

$\{f[g(x)]\}' = f'[g(x)] * g'(x)$

$[x^n]' = n * x^{n-1}$

$[\sin(x)]' = cos(x)$

$[\cos(x)]' = -sin(x)$

$[\tan(x)]' = sec^2(x)$

$[\cot(x)]' = -csc^2(x)$

$[\sec(x)]' = sec(x) * \tan(x)$

$[\csc(x)]' = -csc(x) * \cot(x)$

$[e^x]' = e^x$

$[b^x]' = b^x * \ln(b)$

$[\ln(x)]' = \frac{1}{x}$

$[sin^{-1}(x)]' = \frac{1}{\sqrt{1-x^2}}$

$[cos^{-1}(x)]' = \frac{-1}{\sqrt{1-x^2}}$

$[tan^{-1}(x)]' = \frac{1}{x^2+1}$

$[cot^{-1}(x)]' = \frac{-1}{x^2+1}$

$[sec^{-1}(x)]' = \frac{1}{|x|\sqrt{x^2-1}}$

$[csc^{-1}(x)]' = \frac{-1}{|x|\sqrt{x^2-1}}$

Integration Formulas

$\int dx = x + C$

$\int x^n dx = \frac{x^{n+1}}{n+1} + C$

$\int x^3 dx = \frac{x^4}{4} + C$

$\int a^x dx = \frac{a^x}{\ln(a)} + C$

$\int \frac{dx}{x} = \ln|x| + C$

$\int \cos(x)\, dx = sin(x) + C$

$\int e^x dx = e^x + C$

$\sqrt{x} = x^{\frac{1}{2}}\ so,$

$\int \ln(x)\, dx = xln(x) - x + C$

$\int \sqrt{x} dx = \int x^{\frac{1}{2}} dx = \frac{2x^{\frac{3}{2}}}{3} + C$

$\int \sin(x)\, dx = -\cos(x) + C$

$\int \tan(x)\, dx = -ln|\cos(x)| + C$

$\int \cot(x)\, dx = ln|\sin(x)| + C$

$\int \sec(x)\, dx = ln|\sec(x) + \tan(x)| + C$

$\int \csc(x)\, dx = -ln|\csc(x) + \cot(x)| + C$

$\int sec^2(x) dx = \tan(x) + C$

$\int csc^2(x) dx = -\cot(x) + C$

$\int \sec(x)\tan(x)\, dx = \sec(x) + C$

$\int \csc(x)\cot(x)\, dx = -\csc(x) + C$

$\int \frac{dx}{\sqrt{b^2-x^2}} = sin^{-1}\left(\frac{x}{b}\right) + C$

$\int \frac{dx}{b^2+x^2} = \frac{1}{b} tan^{-1}\left(\frac{x}{b}\right) + C$

$\int \frac{dx}{x\sqrt{x^2-b^2}} = \frac{1}{b} sec^{-1}\left(\frac{|x|}{b}\right) + C$

Some Used Trigonometric Formulas

$Sec(x) = \frac{1}{\cos(x)}$

$\csc(x) = \frac{1}{\sin(x)}$

$sin(x)^2 + cos(x)^2 = 1$

$\sin(x) = \pm\sqrt{1 - \cos(x)^2}$

$\cos(x) = \pm\sqrt{1 - \sin(x)^2}$

$\tan(x)^2 = sec(x)^2 - 1$

APPENDIX

The slope of a tangent line to a curve

Remember the geometric illustration we showed at the beginning of chapter two. Let's suppose we have a circle.

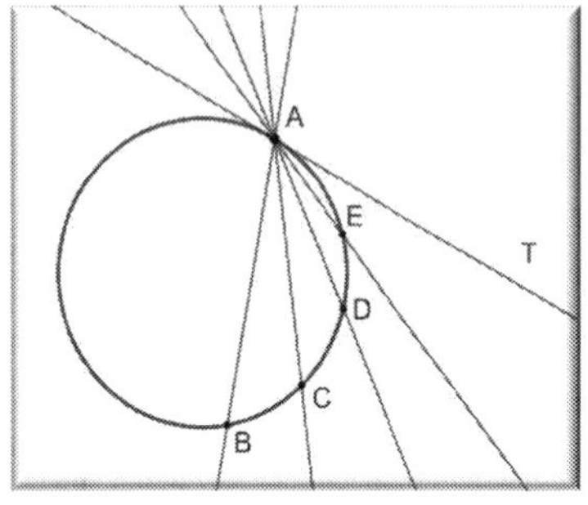

The chord AB is the longest chord compared with the others in this example. As we are approaching the point A going on the circle from point B towards point A, through points B, C, D, E, the length of the chord is becoming smaller and smaller.
When we are at a point which is extremely close to point A, the length of the chord is extremely small going towards zero. The moment we are in point A the chord becomes a tangent to the circle, line AT.

The word tangent comes from the Latin word **tangens** which means "touching". Any tangents to any curve, touch that particular curve in one and only one point.

As we can see below, we have a line that goes through the points S (point 1) and M (point 2). On these points this line intersects the graph of the function:
$f(x) = 4x^3 + 0.7(x-1)^2 + x - 0.5$

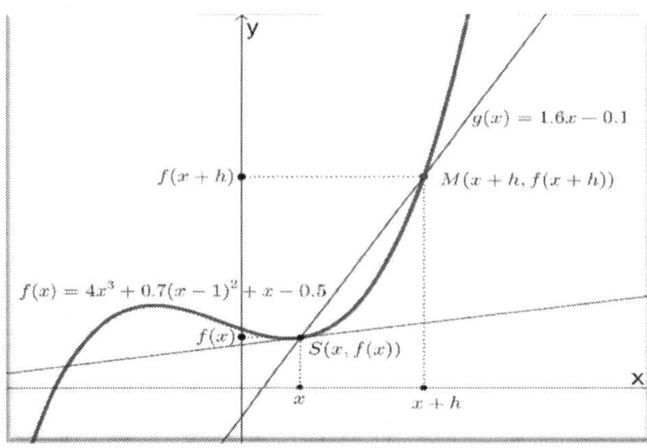

The slope of the line SM can be calculated with the formula:

$$slope\ of\ SM = \frac{y_2 - y_1}{x_2 - x_1} = \frac{f(x+h) - f(x)}{x+h-x} = \frac{f(x+h) - f(x)}{h}$$

Now, we calculate the slope of the tangent that touches the graph of the function f(x) in the point S.

To be able to calculate this tangent we have to consider that the difference between the x coordinates of the points M and S x and x+h respectively, is super small, going towards zero. We need to use the limit when the difference of the coordinates x and x+h goes towards zero.

So,

$x_2 - x_1 = x + h - x = h$

The slope the tangent will be calculated using the limit of the slope formula:

$\lim_{h \to 0} \frac{f(x+h) - f(x)}{h}$

Bibliography

1. Stewart, James – Calculus – Brooks/Cole CENGAGE Learning, Belmont, CA, USA, 2012
2. Woods, David W. – *How Apollo flew to the Moon* – Springer, Glasgow, UK, 2011
3. Freeman, Philip – *Julius Caesar* – Simon & Schuster paperbacks – New York, 2008
4. McCullough, Coleen – *Masters of Rome* series – Century, Simon & Schuster – 1990 – 2007
5. Glancey, Jonathan – *Architecture – A visual history* – Penguin Random House – DK London, 2017
6. https://onekindplanet.org/animal/bear/ fun facts about bears
7. https://www.visitbigsky.com/blog/interesting-facts-about-elk-1/
8. https://thefactfile.org/car-facts/
9. A Brief History of Calculus – https://www.wyzant.com/resources/lessons/math/calculus/introduction/history_of_calculus.

About the Author

Dr. Marcel Sincraian has been working with numbers whole his life, for more than 30 years, as an Engineer, Accountant and Math teacher. While an Engineer, he got his Ph.D. in Civil Engineering. He participated in European Union funded engineering research projects in soil dynamics. As a Math teacher, he noticed that a lot of his students struggle with math and especially with Calculus. Because of his math passion, he decided to help the students with a book in Calculus that is hearth-lighted and interesting. He lives with his family in Burnaby, BC, Canada.

Author, Marcel Sincraian, Ph.D.

Manufactured by Amazon.ca
Bolton, ON

16886152R00118